P9-AGV-133

Microwave Transmission Networks

Planning, Design, and Deployment

Harvey Lehpamer

HL Telecom Consulting, San Diego, California

McGraw-Hill

New York   Chicago   San Francisco   Lisbon   London   Madrid
Mexico City   Milan   New Delhi   San Juan   Seoul
Singapore   Sydney   Toronto

The *McGraw·Hill* Companies

CIP Data is on file with the Library of Congress

Copyright © 2004 by The McGraw-Hill Companies, Inc. All rights reserved. Printed in the United States of America. Except as permitted under the United States Copyright Act of 1976, no part of this publication may be reproduced or distributed in any form or by any means, or stored in a data base or retrieval system, without the prior written permission of the publisher.

1 2 3 4 5 6 7 8 9 0 DOC/DOC 0 1 0 9 8 7 6 5 4

ISBN 0-07-143249-3

The sponsoring editor for this book was Stephen S. Chapman and the production supervisor was Sherri Souffrance. It was set in Century Schoolbook by J. K. Eckert & Company, Inc. The art director for the cover was Margaret Webster-Shapiro.

Printed and bound by RR Donnelley.

McGraw-Hill books are available at special quantity discounts to use as premiums and sales promotions, or for use in corporate training programs. For more information, please write to the Director of Special Sales, McGraw-Hill Professional, Two Penn Plaza, New York, NY 10121-2298. Or contact your local bookstore.

 This book is printed on recycled, acid-free paper containing a minimum of 50% recycled, de-inked fiber.

I dedicate this book to the memory of my late father,
who was a good man and a great engineer.

Contents

Introduction

Microwave radio network design is a subset of activities that constitute the overall transmission network design. Transmission networks are sometimes called *transport networks, access networks,* or *connectivity networks.* For many wireless carriers, microwave is becoming a popular preference over wireline (leased lines) transport for many reasons, especially as microwave radio equipment costs decrease and installation becomes simpler. Low monthly operating costs can undercut those of typical single (and especially multiple) T1/E1 expenses, proving it to be more economical over the long term—usually two to four years. Network operators also like the fact that they can own and control microwave radio networks instead of relying on other service providers for network components.

Most people in the telecommunications field, especially transmission engineers, project managers, and network planners in wireless systems, should have at least a basic understanding of the planning, design, and deployment process of the microwave network.

For clarity and technical correctness, we should be very clear and consistent in the terminology used throughout this book. It is important to remember that not all microwave systems are point-to-point, and not all point-to-point systems are microwave. Although many principles are common to other microwave systems, this book predominantly deals with the terrestrial microwave point-to-point systems in 2 to 60 GHz.

This book covers all stages of terrestrial microwave point-to-point network build-out from initial planning and feasibility studies to real system deployment. Emphasis is given to practical guidelines and activities involved in putting microwave system into operation. It describes the process behind planning and creating a business case for a microwave network, including the advantages and disadvantages, and includes discussions that will help executives to make an informed decision about whether to build a microwave network.

What is a difference between planning, design, and deployment? Although distinct differences exist in telecommunications projects, these three activities in microwave network build-out are somewhat overlapping and mutually dependent. Many times, partial design or redesign has to be performed during the planning stage and/or deployment as well.

Planning usually refers to a high-level decision-making process that encompasses budget and schedule definition and identifying team members required for the project. It also includes determination of frequency band, system capacity, network configuration, and performance objectives.

System design is an actual detailed link engineering process (which may or may not include site visits) that includes creation of the detailed bill of materials, ordering equipment (MW radio, shelters, towers, and other transmission hardware and software), ordering engineering, installation and other services, and so forth.

Deployment (also called *implementation*) includes all of the field activities such as site and path surveys, tower erection, equipment installation, creation of an as-built documentation, and acceptance testing and commissioning.

Details and mathematical models of microwave point-to-point link engineering will be discussed only briefly, as they are beyond the scope of this book and its intended audience. In addition, literature and standards are available that provide more details. A comparison of North American and ITU microwave design models and methods is presented, and project management and logistics issues, deployments in different countries, and regulatory and ethical issues are discussed in more detail.

This book will be a useful source of information for project managers, sales executives, and nontechnical personnel involved in planning and/or making decisions about microwave network deployment. It will also be helpful to managers and directors of engineering and operations who work for carriers or wireless operators. Additionally, it will serve engineers who are starting their careers in microwave field, engineers whose main field is not RF or network planning, telecom and microwave engineers who want to expand their knowledge about the business side of microwave network build-out, and instructors and consultants in the telecommunications field—and anyone else who is involved in real-life microwave network build-out.

For those who want only an overview of the microwave network build-out, Chapter 7, "Project Management," will provide easy reading and, at the same time, sufficient high-level information to get them started. More detailed information can be found in other chapters of the book.

An extensive glossary provides definitions of many commonly used terms in transmission, RF, and microwave fields. An understanding of these terms is necessary to comprehend the material in this book.

Harvey Lehpamer

1

Transmission Network Fundamentals

1.1 Transmission Network Media

In telecommunications, information can be transmitted between two locations using a signal that can be either analog or digital in nature. In the telecommunications networks today, digital transmission is used almost exclusively, in which analog traffic, such as voice calls, is converted to digital signals (a process referred to as *sampling*) to facilitate long distance transmission and switching.

A high-pitched voice contains mostly high frequencies, while low-pitched voice contains low frequencies. A loud voice contains a high-amplitude signal, while soft voice contains a low-amplitude signal. Analog signals can be combined (i.e., multiplexed) by combining them with a carrier frequency. When there is more than one channel, this is called *frequency-division multiplexing (FDM)*. FDM was used extensively in the past but now has generally been replaced with the digital equivalent, called *time-division multiplexing (TDM)*. The most popular TDM system is known as the Tier 1 (T1) system, in which an analog voice channel is sampled 8,000 times per second, and each sample is encoded into a 7-bit byte. Twenty-four such channels are mixed on two copper pairs and transmitted at a bit rate of 1.544 megabits per second (Mb/sec). T1 in North America (E1 in the rest of the world) remains an important method of transmitting voice and data in the *public switched telephone network (PSTN)*.[1]

A talking path (i.e., a switched circuit) in the PSTN can be either analog or digital or a combination thereof. In fact, a digital signal can be transmitted over a packet-switched network as easily as a circuit-switched network. Digitized voice is little different from data; therefore, if data can be transmitted over a packet network, then so can digitized voice. One of the most common applications is now known as *voice over Internet (VOI)*. The challenge, of course, is to get the transmitted signal to the destination fast enough (delay-related issues), as in instances in which the conversation may be time sensitive. A second challenge is to get each packet, which is a small piece of a voice conversation, to its destination in the proper order.

Three types of media (physical layer) can be used in transmitting information in the telecommunications world:

- Copper lines (twisted-pair and coaxial cables)—for low- and medium-capacity transmission over a short distance,

- Fiber-optic transmission—for medium- and high-capacity transmission over any distance,

- Wireless transmission

  - Low- (mobile radio) and medium-capacity (microwave point-to-point) over short and medium distances

  - Satellite—for low- and medium-capacity transmission over long distances

The terms *transmission* and *transport* are used interchangeably in this text, as both are currently in use; the former is preferred in Europe, but the latter is more commonly used in North America. Sometimes *transmission* refers only to the physical media, while *transport* can include other OSI Model layers of the data transfer.

1.1.1 Wireline Systems

Years ago, copper wire was the only means of transporting information. Technically known as *unshielded twisted pair (UTP)*, it consists of a large number of pairs of copper wire of varying size within a cable. The cable did not have a shield, so the signal (primarily the high-frequency part of the signal) was able to leak out. In addition, the twisting on the copper pair was very casual, designed as much to identify which wires belonged to a pair as to handle transmission problems. Even with these limitations, it was quite satisfactory for use in voice communications.

Coaxial cable technologies were primarily developed for the cable TV industry. In the last few years, this technology has been extended to provide Internet services to residences. The high capacity of coaxial cable allows it to support multiple TV channels, and this capacity can also be used for high-speed Internet access. Like fiber optics, the cost of cable installation limits the deployment of new services, and current deployments are not typically in areas that allow this service to be offered to business offices.

Fiber optics constitute the third transmission medium, and this is unquestionably the high-bandwidth transmission medium of choice today. Fiber-optic cables can be placed in ducts, buried in the ground, suspended in the air between poles, installed as part of the ground wire on the high-voltage transmission towers *optical power ground wire* (OPGW), and so forth. Transmission speeds of as high as 10 Gbps have become commonplace in the industry. Of course, laying fiber, on a per-mile basis, still costs somewhat more than laying copper, but on a per-circuit basis there is no doubt that fiber is more cost effective. The huge capacity of fiber certainly makes for more efficient communications; however, placing so much traffic on a single strand, for point-to-point communications, makes for greater vulnerability. Most of the disruptions in the long distance network are a result of physical interruption of a fiber run (called *backhoe fade*), and the ring configuration is the protection solution used most often in fiber-optic networks.

The cost of laying fiber-optic cable can easily reach $70,000 to $150,000 per mile, excluding terminal equipment. For that reason, most users opt for either leasing fiber-optic facilities or building their own microwave network.

1.1.2 Wireless Systems

Wireless communications can take several forms: microwave (point-to-point or point-to-multipoint), synchronous satellites, low Earth-orbit satellites (LEOs), cellular, personal communications service (PCS), and so on. For years, microwave radio transmissions have been used in the telecommunications industry for the transport of point-to-point data where information transmissions occur through carrier signals. Microwave carrier signals are typically relatively short in wavelength and can transmit information through various modulation methods.

To understand wireless technology, a basic understanding of the radio frequency (RF) spectrum is required. The RF spectrum is a part of the electromagnetic spectrum in which a variety of commonly used devices (including television, AM and FM radios, microwave radios, cell phones, pagers, and many other devices) operate. The electromagnetic spectrum has been used for communications for over 100 years, and it comprises an infinite number of frequencies, e.g., such as AM radio at 1 MHz or the cellular/PCS band at 2 GHz. Frequencies are measured in cycles per second, or hertz, which are inversely related to wavelength. At low frequencies, wavelengths are long; at higher frequencies, wavelengths are very short. Given an equal power level, the longer the wavelength, the greater the distance the signal can travel.

Whereas low-frequency signals (such as AM radio) can be transmitted for hundreds of miles, high-frequency signals (such as infrared) can travel only a few feet.

Microwave and millimeter-wave bands occupy frequencies from around 1 to 300 GHz, but this book discusses the characteristics of the bands for terrestrial microwave systems from around 2 to 60 GHz.

The RF spectrum in which these carrier transmissions occur is subject to regulation in the United States by the Federal Communications Commission (FCC), Industry Canada in Canada, and globally via the International Telecommunications Union. Countries that are members of the ITU generally follow the ITU spectrum allocation.

Within the RF spectrum, not all frequencies are subject to licensing requirements, and license-free bands include the industrial, scientific, and medical (ISM) band (the most widely used license-free frequency band) and the Unlicensed National Information Infrastructure (U-NII) Band.

1.1.3 Alternatives

Free-space laser communications systems are wireless connections through the atmosphere that employ the optical part of the frequency spectrum. Therefore, they cannot be categorized as either wireless or wireline systems in a classical sense. They work only under clear line-of-sight conditions between each unit, eliminating the need for securing right of ways, buried cable installations, and government licensing. Free-space laser communications systems can be quickly deployed, since they are small and do not need any radio interference studies. Optical wireless is an attractive option for multigigabit-per-second short range links (typically from a few hundred meters up to 2 km) where laying optical fiber is too expensive or impractical, and where microwave systems do not provide enough bandwidth.

This type of optical communication, also known as free-space optical (FSO), has emerged as a commercially viable alternative to RF and millimeter-wave wireless for reliable and rapid deployment of data and voice networks. (See Fig. 1.1, which shows wireless transceivers mounted on rooftops.) RF and millimeter-wave technologies allow rapid deployment of wireless networks with data rates from tens of megabits per second (point-to-multipoint) up to several hundred megabits per second (point-to-point). However, spectrum-licensing issues at licensed and interference at unlicensed frequency bands can sometimes still limit their market penetration.

Although optical losses and space losses can introduce a significant attenuation of the optical signal, the main problem is atmospheric effects. As mentioned earlier, it is desirable to have as much excess mar-

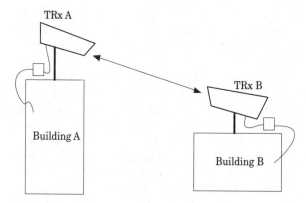

Figure 1.1 Optical wireless transceivers.

gin as possible to mitigate atmospheric effects such as fog. On a sunny day, the atmosphere is clear, and the margin is useful to overcome fades caused by turbulence. On a foggy day, the margin is used to overcome signal attenuation. The dominant atmospheric effect that affects optical communication is attenuation of the signal by scatter and absorption. Molecular scatter and absorption of major atmospheric constituents is relatively insignificant. Although rain and snow can cause attenuation up to approximately 40 dB/km and 100 dB/km, respectively, fog by far is the largest problem. In extremely heavy fog, attenuation as high as 300 dB/km (500 dB/mi) has been reported. Clearly, either link distance or link availability is compromised as part of the network design.

Alignment is important both at the transmitter and the receiver. The transmitter has to be pointed accurately to ensure efficient delivery of energy to the receiver. The receiver has to be pointed properly to ensure that the signal entering the receiver aperture makes it to the detector. A great deal depends on how and where the transceivers are located and, in the case when a transceiver is mounted on the roof of a tall building, building sway contributes significantly to pointing error. A high-rise building can sway more than a meter, and the roof itself often houses air-conditioning and ventilation units as well as elevator and other mechanisms that cause vibration in the low tens of hertz. Moreover, some roofs are not very solid and can deform when someone walks on them.

Temperature changes (diurnal and seasonal) and uneven heating by the sun can deform the mount enough to throw the pointing off. Some FSO users have indicated that they need to realign their transceiver units several times a year for this very reason. Optical wireless transceivers and their mounts have enough wind resistance that they can be tilted in heavy winds. To make things worse, building owners, for liability reasons, do not readily grant approval for the installation of penetrating mounts, and nonpenetrating rooftop mounts exhibit even greater pointing fluctuations.

Microwave and infrared transmission systems are both good alternatives for short distance network connections (less than 2 mi). Each system requires line-of-sight and has its advantages and disadvantages. The ease of obtaining frequency spectrum licensing, various weather and atmospheric conditions, topology of the area, and security should be evaluated to determine which system is used. When frequency spectrum licensing is difficult and expensive to obtain, infrared transmission systems have a distinct advantage. Infrared systems are also advantageous if the weather is normally rainy, but not foggy, and there is little smog; when the conditions require spanning a large body of water; and when the area has high levels of electromagnetic interference (EMI). In addition, an infrared transmission system is less inclined to be intercepted.

Microwave transmission systems are advantageous in areas that are foggy and have a substantial amount of snow or smog, or if the conditions require spanning longer distances. Microwave systems should be used anytime the distance exceeds two miles.

1.2 Basic Terminology

1.2.1 E1 and T1

There are two standards for the first-order digital transmission systems. The T1 system, developed by Bell Laboratories, is used mainly in the U.S.A., Canada, Taiwan, Jamaica, and few other countries. North American T1 service providers often refer to the signal interfaces between the user and the network as "DS1" signals. In the case of user-to-user interfaces, the term "DSX-1" is used to describe those DS1 signals at the "cross-connect" point; therefore, DSX-1 is a physical interface of the T1 circuit. Most of the countries around the world use E1 system defined by European Conference of Postal and Telecommunications Administration (CEPT).

Details of the T1 systems can be found in literature.[2] Here, only a brief description and a definition of T1 are given:

- There are 24 DS0 (64-kbps) time slots in a T1 line, providing a total bandwidth of 24×64 kbps = 1,536,000 bps (1,536 kbps).

- Another 8,000 bps are used as framing bits.

- Adding up the total bandwidth of the 24 DS0 channels and the 8-kbps framing bits yields a 1,544 kbps T1 data rate.

The North American digital hierarchy (Table 1.1) starts with a basic digital signal level of 64 kbps (DS0). Thereafter, all facility types are usually referred to as "Tx," where "x" is the digital signal level within the hierarchy (e.g., T1 refers to the DS1 rate of 1.544 Mbps). Up to the DS3 rate, these signals are usually delivered from the provider on twisted-pair or coaxial cables.

TABLE 1.1   North American Data Rates

Name	Rate (kbps)
DS0	64
DS1	1.544
DS1C	3.152
DS2	6.312
DS3	44.736
DS4	274.176

The basic format for T1 transmission facilities in Japan (Table 1.2) is similar to North American ANSI standard and called J1. However, the CMI line coding used in Japan is different from the one used in North America.[3] Rather than using 50 percent duty cycle (half-width) "mark" pulses, continuous marks are transmitted, inverting the voltage after each mark. The spaces also have a transition, but in the middle of the pulse period. This ensures that there are always sufficient transitions to maintain synchronization, regardless of whether the signal is all zeroes or all ones.

TABLE 1.2   Japanese Digital Hierarchy

Designation	Bit rate (Mbps)	Number of voice channels
DS1	1.544	24
DS2	6.132	96
DS3	32.064	480
DS4	97.728	1440
DS5	400.352	5760

The CCITT Digital hierarchy's[4] basic level is the DS0 rate of 64 kbps (Table 1.3). These signals are usually delivered from the pro-

TABLE 1.3   ITU Data Rates

Name	Rate (kbps)
DS0	64
E1	2.048
E2	8.448
E3	34.368
D4	139.264

vider via twisted-pair or coaxial cables. The following is a description of "CEPT" digital hierarchy:

- There are 30 DS0 (64-kbps) time slots in an E1 line, providing a total bandwidth of 30 × 64 kbps = 1,920,000 bps (1,920 kbps).
- One 64-kbps time slot (TS0) is used for framing bits.
- Another 64-kbps time slot (TS16) is used for signaling of voice frequency channels.
- Adding up the total bandwidth of the 30 DS0 channels, framing, and signaling bits yields the 2,048 kbps E1 data rate.

1.2.2   PDH, SDH, and SONET

Traditionally, transmission systems have been asynchronous, with each terminal in the network running on its own clock. In digital systems, clocking (timing) is one of the most important considerations. Timing means using a series of repetitive pulses to keep the bit rate of the data stream constant and to indicate where the ones and zeroes are located in a data stream. Because these clocks are free running and not synchronized, large variations occur in the clock rate and thus the signal bit rate.

Asynchronous multiplexing uses multiple stages; lower-rate signals are multiplexed, and extra bits are added (bit-stuffing) to account for the variations of each individual stream and combined with other bits (framing bits) to form higher-level bit rates. Then bit-stuffing is used again to produce even higher bit rates. At the higher asynchronous rate, it is impossible to access these signals without multiplexing.

The *Plesiochronous Digital Hierarchy (PDH)* signals have the essential characteristics of time scales or signals such that their corresponding significant instants occur at nominally the same rate. The prefix "plesio," which is of Greek origin, means "almost equal but not exactly," meaning that the higher levels in the CCITT (ITU today) hierarchy are **not** an **exact** multiple of the lower level. Any variation in rate is constrained within specified limits. The PDH systems belong to the first generation of digital terrestrial telecommunication systems in commercial use.

Synchronous network hierarchies were introduced in the late 1980s. The term "synchronous" means occurring at regular intervals and is usually used to describe communications in which data can be transmitted in a steady stream rather than intermittently. For example, if a telephone conversation were synchronous, each party would be required to wait a specified interval before speaking.

Synchronous Digital Hierarchy (SDH) is a newer technology in the field of digital transmission.[5] The transmission is carried out in a synchronous mode, hence the name. The most important advantage in adopting this synchronous technology is to enable the mapping of various user bit rates directly onto the main transmission signals, thus bypassing various stages of multiplexing and demultiplexing as was done in the case of earlier PDH technology.

North American *Synchronous Optical Network (SONET)* is a second-generation digital optical transport protocol. The fiber-based carrier network uses synchronous operations among such network components as multiplexers, terminals, and switches.

The number of new architectures and topologies are made possible as a result of this new technology. For instance, add/drop multiplexers (ADMs) have made it possible to use SONET/SDH terminals in a long chain with bit streams added or dropped along the way in an effective manner.[6] By closing the chain at two ends, a ring configuration is possible, which provides enhanced protection features.[7,8] SONET-based rings create a robust, high-availability network that can "heal" itself automatically by routing around failures. The optical fiber rings offer security, high bandwidth, low signal distortion, and high reliability.

Other advantages include support of network management systems, easy upgrade to high bit rates, and adaptability to the existing PDH. In synchronous networks, all multiplex functions operate using clocks derived from a common source.

The North American SONET system is based on multiples of a fundamental rate of 51.840 Mbps, called STS-1 (for synchronous transmission signal, level 1). The facility designators are similar but indicate the facility type, which is usually fiber-optic cable (e.g., OC-1 is an optical carrier supporting an STS-1 signal, whereas OC-3 supports a STS-3 signal, and so forth). Some typical rates are listed in Table 1.4.

The International *Synchronous Digital Hierarchy (SDH)* system is based on a fundamental rate of 155.520 Mbps, three times that of the SONET system. This fundamental signal is called STM-1 (synchro-

TABLE 1.4 STS Data Rates

Name	Rate (Mbps)
STS-1	51.840
STS-3	155.520
STS-9	466.560
STS-12	622.080
STS-48	2488.320

nous transmission module, level 1). The typical transmission media is defined to be fiber, but the broadband ISDN specification does define a *user-network interface (UNI)* STM-1 (155.520 Mbps) operating over co-axial cables. Some typical rates within this hierarchy are shown in Table 1.5.

TABLE 1.5 STM Data Rates

Name	Rate (Mbps)
STM-1	155.520
STM-3	466.560
STM-4	622.080
STM-16	2488.320

One of the main advantages of the ITU-T SDH system is the fact that it may be the first compatible system used worldwide. A further advantage is the extremely high bit rate transmitted by the system (e.g., nearly 10 Gbps with STM-64). When used in conjunction with Dense Wavelength Division Multiplexing (DWDM), even much higher rates can be handled. SDH is perfectly suitable to multiplex and transport the traffic of PDH networks, and it also can be used for data transport and for leased lines. It follows further from the synchronizing feature that a low-level container, also including its content, can be accessed at any higher hierarchical level. However, a disadvantage of SDH is the necessity to establish a synchronization network.

SDH and SONET are both gradually replacing the higher order of PDH systems, but the evolution from PDH to SDH worldwide is a phased process because SDH based networks must have the flexibility to utilize existing PDH transport media. The advantages of higher bandwidth, greater flexibility, and scalability make these standards ideal for asynchronous transfer mode (ATM) networks as well.

To maximize the benefits of the SONET/SDH microwave radio, the radio must be capable of complementing a synchronous fiber-optic network. This means that, for the microwave radio to integrate with fiber-optic network elements, its design must address a number of parameters, including capacity and growth, network management, maintaining pace with SDH/SONET standards evolution, interface, and performance. Providing microwave radio with the optical interface will allow a microwave network to integrate with the fiber-optic network without the use of multiplexing equipment (unless drop and/ or insert of the traffic is required). The same management tools are used for both media. Common expectations for fiber *and* microwave elements are signal rates and interfaces, overhead processing, service channels, operations systems, and transmission quality. SONET/SDH microwave radios can easily integrate into a new or existing fiber-optic network (see Fig. 1.2).

The new generation of digital microwave systems, based on synchronous digital hierarchy (SONET/SDH), can meet the requirements for the high-capacity backbone transmission systems. SDH/SONET radios provide an economical solution when existing infrastructure (towers, shelters, and so on) can be reused, and when rights of way or adverse terrain make fiber deployment very costly or time consuming.

Today's SONET/SDH radio technology is capable of delivering bandwidth-efficient bandwidth of 8 bits/s/Hz. For example, 512-state quadrature amplitude modulation (QAM) technique can pack two STM-1 streams into a single 40-MHz channel using a single carrier. By adding channels in a 1:N configuration, system capacities of up to 14 protected STM-1s can be achieved within one frequency band (for example, in the upper 6-GHz band, 8 bidirectional channels are available). By deploying a dual-band configuration (such as lower 4- and 5-

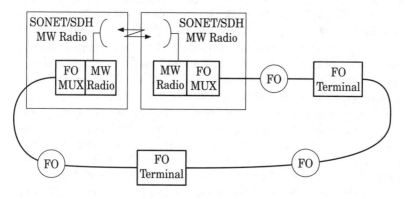

Figure 1.2 Hybrid microwave/fiber optic ring.

GHz bands) system, capacities of STM-16 and greater are achievable. Although theoretically possible, this can be done only if the spectrum governing bodies allow more than one channel to be accessed by the same user.

1.2.3 ATM

Asynchronous transfer mode (ATM) is the complement of *synchronous transfer mode (STM)*, which we discussed earlier. STM is a circuit-switched networking mechanism whereby a connection is established between two termination points before data transfer commences and torn down when it is completed. In this way, the termination points allocate and reserve the connection bandwidth for the entire duration, even when not actually transmitting data.

ATM is a transmission technology that uses fixed-size packets called *cells*. A cell is a 53-byte packet with 5 bytes of header/descriptor and 48 bytes of payload, or user traffic (voice, data, video, or their combination). Today, telecommunications companies are deploying fiber optics in cross-country and cross-oceanic links with Gbps speeds. They would like to carry, in an integrated way, both real-time traffic such as voice and high-resolution video, which can tolerate some loss but not delay, as well as non-real-time traffic such as computer data and file transfer, which may tolerate some delay, but not loss.

The problem with carrying these different characteristics of traffic on the same medium in an integrated fashion is that the requirements of these traffic sources may be quite different. In other words, the data comes in bursts and must be transmitted at the peak rate of the burst, but the average arrival time between bursts may be quite large and randomly distributed. For these connections, it would be a considerable waste of bandwidth to reserve a bucket for them at their peak bandwidth rate for all times when, on the average, only 1 in 10 buckets may actually carry the data. Thus, using the STM mode of transfer becomes inefficient as the peak bandwidth of the link, peak transfer rate of the traffic, and overall burstyness of the traffic (expressed as a ratio of peak/average) all go up. Terms such as *fast packet, cell,* and *bucket* are used interchangeably in ATM literature. ATM networks are connection-oriented packet-switching networks.

Future telecommunication networks, including wireless networks, must be able to offer today's range of services as well as services with new features; e.g., variable bit rates. The requirements of modern networking involve handling multiple types of traffic (voice, video, and data), all with individual characteristics that make very different (and often opposed) demands on the telecommunication channel. The second requirement is reliability and flexibility in the communication links. The greatest problem is that transmissions occur at statistically random intervals with variable data rates. A way of solving this problem is to use a service that takes packets on the transport layer from a higher layer and fragments them in small packets of a fixed size. The delays produced by each packet are going to be short and probably fixed, so, if voice and video traffic can be assured priority handling, they can be mixed with data without diminishing any reception quality. The services are called the *ATM adaptation layer (AAL)*, and the packets are called *cells*. A new adaptation layer was required to provide the flexibility for network operators to control delay on voice services and to overcome the excessive bandwidth needed by using structured circuit emulation.[9] The new AAL2 was designed specifically for cost-effective voice transport. AAL2 is used in 3G wireless networks as a backhaul connection between *radio base stations (RBSs)* and *base station controllers (BSCs)*.

Over the last few years, more and more transmission systems, especially those in wireless networks, have been using ATM over the microwave networks. These radio systems carrying packetized traffic (ATM or frame relay) have to be designed in a way that takes into account the behavior of this kind of traffic. Because ATM is primarily designed for an essentially error-free environment, in the wireless arena, the sources of errors and their consequences on ATM traffic and its quality of service (QoS) are being studied today. ATM traffic requires a very high-quality transmission medium with a good background error rate (also called *residual BER* or *RBER*). Microwave radio and fiber optics are ideal from this perspective, as they both offer in the order of 10^{-13} background error rates. Microwave radio is subject to fading for a small period of time but, by proper design, this is limited to a specified time period (typically better than 99.99 percent availability).

ATM is designed for low BER links, and radio links with a moderate BER can cause unacceptable high cell loss and misinsertion rates.[10] By definition, a misinserted cell is a received cell that has no corresponding transmitted cell on the considered connection. Cell misinsertion on a particular connection is caused by defects on the physical layer affecting any cells that were not previously associated with this connection.

1.3 Transmission Network Topology

The main objectives of the transmission network are to connect all the points of interest, satisfy the capacity demands, and provide reliable service using different media (microwave, copper, fiber optics, or satellites). During the transmission network build-out, it is always important to establish a transmission network plan that will include all present traffic requirements as well as future expansion.

Network operators worry about two things: how to start deploying the network in phases without spending capital before it is needed, and how to grow these small, initial segments once they see growth coming down the road. That is why scalability is high on everyone's list. The transmission network topologies can be divided into two groups.

- The *flat network* is a single entity, which means that it can be optimized to have high network utilization. A drawback is that the network topology is sensitive to changes in the traffic distribution. A change in the distribution changes the earlier, well optimized network to being nonoptimized.

- The *layered network* facilitates the design of a network and its subsequent expansion, as the total network is modular. Modular design also makes the network and the traffic routing easy to understand, thus simplifying operation and maintenance, which will reduce the operator's operations and maintenance (O&M) costs in the future.

A layered network is divided into various network layers, which are connected via gateways. The layered/modular network is designed subnetwork by subnetwork; i.e., the total demand matrix is divided into demand matrixes for each subnetwork. The new, smaller matrixes are easier to handle and to understand than are the large ones.

In the future, as the services offered to the end user become more and more flexible, the layered approach might be the most suitable topology—assuming that the initial cost is not considered.

In wireless networks, the size of the network is assessed based on the number of cell sites and/or required backhaul capacity. Many successful mobile operators protect transmission by using automatic traffic rerouting, assuring additional reliability in normal situations, such as when microwave radio access links suffer cut-off as a result of poor weather conditions, possible fiber-optic cable cuts, or any other human error. With a flexible rerouting transmission system, such as T1/E1 trunk rerouting, backup capacity can pass via physically separate routes, given that the problem is not likely to interrupt both routes simultaneously.

Rerouting can be arranged for all sites or only critical sites, such as base stations that are labeled as higher priority—for example, a hub site. Hub sites are those sites that collect traffic from more than one other site (typically three to four other sites) and carry that traffic toward the BCS location or fiber-optic ring hub site.

For a larger transmission network, it is recommended that one use a ring configuration as a high-capacity backbone that carries traffic to the switch. The ring architecture is considered to be a reliable communication facility, as it provides automatic protection against the following:

- Site hardware (batteries, towers, antenna systems) failures
- Radio/mux equipment failures
- Propagation failures in the microwave network
- Cable cuts in the fiber-optic network

It also provides basic user features such as simple operation, fault location, and maintenance. Ring configuration automatically provides alternative routing of E1/T1 traffic and no loss of E1/T1 traffic due to a single failure. Each E1/T1 circuit must be dedicated completely around the ring, and reuse of same E1/T1 in the opposite direction is not possible. For ultimate reliability, both directions can be "1+1" hardware protected.

In PDH networks, additional hardware with built-in intelligence to assess the T1/E1 quality and switch circuits will be required. This hardware has to be added at every site, and it is useful for small networks. SONET/SDH have incorporated several protection/switching techniques from their inception. These include linear APS, path-switched rings, line-switched rings, and virtual rings. These techniques provide the ability for a network to detect the problem (under 10 ms) and heal itself automatically in the case of failure, with the restoration time less than 50 ms. Self-healing schemes use fully duplicated transmission systems and capacity for alternate routing of today's time division multiplexed (TDM) or synchronous transfer mode (STM) circuit facilities. The restoration capacity and the associated transmission systems are essentially unused except in the rare occasions of network failure.

Although expensive and relatively complex to implement, the dual-homed ring architecture is the choice for high-capacity digital service providers. This architecture uses a drop-and-continue feature that ensures that traffic is available to pass between adjacent rings at two separate nodes. If an entire node is lost, the receiving ring equipment will select traffic from the other node. Although it looks expensive, actually network survivability has a high potential for cost reduction in the future.

1.4 Transmission Network Performance

In today's networks, with converged voice and data, performance degradation may be as dangerous and costly as hardware failures. Degraded transmission network can result in unacceptable signal transmission quality, loss of information, and dropped connections. High availability does not mean just preventing catastrophic failures; it also means preventing quality and performance degradation.

High availability of the transmission network is an end-to-end network goal. A network management system (NMS) can help identify critical resources, traffic patterns, and performance levels. Transmission network survivability is usually measured in terms of its long-term availability or average network uptime. Most operators expect their network to be continuously available (or at least with as little downtime as possible) to minimize potential loss of revenue. The survivable network has an infrastructure of transmission facilities and reliable network elements that are used to manage them. High network availability at the transport level may be achieved using millisecond restoration schemes provided by self-healing network configurations such as SONET/SDH rings or *fast facility protection (FFP)*. *Digital access cross-connects (DACS)* in combination with SONET/SDH ring configurations will ensure network availability and survivability. An FFP network comprises two physically diverse routes with identical transmission systems (route diversity). Each route carries half of the working traffic and half of the restoration traffic. The restoration traffic on each route is the duplicate of working traffic on the other route. If the media on these routes are different (i.e., one is fiber optic and the other one is, for example, microwave), we talk about *media diversity*. These highly reliable solutions do not come cheap and, in many cases, a compromise between the cost of the network and its deployment time and reliability has to be made.

Regardless of the transmission network medium and topology, hardware redundancy is an option when designing the transmission network. Protection types usually employed are "1+1," where one card or module serves as a protection for another one, or "1+N," where one card or module protects N other units. Linear "1+1" protection switching means that identical payloads will be transmitted on the working and protect fibers, or working and protect frequencies in case of the microwave system. Linear "1+N" protection switching assumes the existence of one protect fiber/frequency for N working fibers/frequencies.

A rule of thumb is that, if all the hardware is protected with a "1+1" and/or "1+N" configuration, fewer spare parts are needed. In the case of hardware failure, protection will kick in, and the operator will have sufficient time (which may be days or weeks) to order replacement parts from the supplier. In addition, the ring configuration could provide protection against hardware failures as well, so additional hardware protection might not be required. This is something that transmission engineers have to decide, and the decision will be based on both technical and budgetary requirements.

It is also important to remember the distinction between terms like *performance, availability, quality,* and so on. Although the network is operational (available), network performance can still be poor with increased levels of BER, for a number of reasons. In voice networks, that may mean reduced sound quality; in data networks, it can cause constant data retransmissions, incorrect information, and so forth. In other words, the quality of such a network is low and unsatisfactory.

1.5 Network Synchronization

Digital systems require all signals within a given transmission level to maintain a frequency relationship. If this relationship is not maintained, information will be lost, or transmission capacity will be underutilized. In addition, varying transmission times, inaccurate timing, and unstable network equipment can also cause bits traveling in a transmission path to arrive at a time that is in variance with their expected arrival. Solving timing-related problems is part of network synchronization. To date, two approaches have been dominant in achieving network synchronization: *plesiochronous,* used in PDH networks, and *synchronous,* used in SDH networks.

Digital network connectivity depends on the availability of a reliable synchronization source to provide a timing reference to the network elements. Synchronization networks provide timing signals to all synchronization network elements at each node in a digital network. Buffer elements at important transmission interfaces absorb differences between the average local frequency and the actual short-term frequency of incoming signals, which may be affected by phase wander and jitter accumulated along the transmission paths.[11]

A *slip* in T1 is defined as a 1-frame (193 bits) shift in time difference between the two signals in question. This time difference is equal to 125 µs, and these slips are not a major impairment to trunks carrying voice circuits. The lost frames and temporary loss of frame synchronization results in occasional pops and clicks being heard during a call in progress. However, with advancements in DS1 connectivity, these impairments tend to spread throughout the network. To minimize them, a hierarchical clock scheme was developed whose function was to produce a primary reference for distribution to switching centers so as to time the toll switches.

Stratum 1 is defined as a completely autonomous source of timing that has no other input other than perhaps a yearly calibration. The usual source of Stratum 1 timing is an atomic standard or reference oscillator. The minimum adjustable range and maximum drift is defined as a fractional frequency offset of 1×10^{-11} or less. At this minimum accuracy, a properly calibrated source will provide bit-stream timing that will not slip relative to an absolute or perfect standard more than once every 4 to 5 months. Atomic standards, such as cesium clocks, have far better performance. A Stratum 1 clock is an example of a *primary reference source (PRS)* as defined in ANSI/T1.101. Alternatively, a PRS source can be a clock system employing direct control from *coordinated universal time (UTC)* frequency and time services such as *global positioning system (GPS)* navigational systems. The GPS system may be used to provide high-accuracy, low-cost timing of Stratum 1 quality.

A *Stratum 2* clock system tracks an input under normal operating conditions and holds to the last best estimate of the input reference frequency during impaired operating conditions. A Stratum 2 clock system provides a frame slip rate of about approximately one slip in seven days when in the hold mode.

Stratum 3 is defined as a clock system that tracks an input as in Stratum 2 but over a wider range. Sometimes Stratum 3 clock equipment is not adequate to time SONET network elements. Stratum 3E, which is defined in Bellcore documents, is a new standard created as a result of SONET equipment requirements.

Stratum 4 is defined as a clock system that tracks an input as in Stratum 2 or 3 but has the wider adjustment and drift range. In addition, a Stratum 4 clock has no holdover capability and, in the absence of a reference, free runs within the adjustment range limits.

Any Stratum clock will always control strata of lower-level clocks. Inadequate timing may produce problems in any digital network, so the objectives have to be set very early in the planning process. The master timing sources that are possible to use and recommended are as follows:

- PRS DS1—Timing is received from a collocated primary reference source such as a GPS or LORAN-C receiver, which is Stratum 1 level.

- Dedicated DS1—Timing is obtained by terminating a DS1 dedicated to synchronization.

- Traffic-carrying DS1—Timing is extracted from a traffic-carrying DS1 (PDH or copper medium and not carried over from SDH/SONET) coming into a site.

- SDH—Timing is extracted from the SDH line signal where there is no influence from the pointer movements phase steps.

- SONET DS1—Timing is obtained by deriving a non-traffic-carrying DS1 from SONET network elements, an OC-N MUX.

Telecommunications equipment (for example, cellular base stations) receive their timing signals from the worldwide GPS, an array of satellites that beam timing signals accurate to within 300 nsec to receiving stations located around the globe. If the link fails, a backup system must be in place to maintain timing accuracy; if there is no backup, the cell goes down, and communication is lost, which is an unacceptable condition. The GPS time standard is downlinked from the satellite as long as it is in range of the GPS receiver. However, there are intervals of time when a satellite is not in range (overhead), and no GPS time standard is available. This period without any GPS time update is called the *GPS holdover time.*

One major problem encountered after designing a timing network is evaluating its performance, as any problems related to synchronization can be difficult to detect and even more difficult to troubleshoot.

The digital microwave radios and high-level multiplexers are not considered in the synchronization plan, since they internally synchronize on a per-hop basis. New synchronous radios on the market provide a transparent transmission media, which in SDH terms runs in a default *regenerator section termination mode (RST)*. In this mode, the radio obtains synchronization from the aggregate 155 Mbps input into the radio, and they do not offer or require any additional synchronization options.

1.6 Network Delays

Signal *propagation delay* or *latency* describes the delay of a transmission from the time it enters the network until the time it leaves; *low latency* means short delays, whereas *high latency* means long delays. Low latency is essential for real-time transmissions. These include live voice conversations (but not voice mail messages, which are time insensitive) and live two-way video (but not entertainment video clips, which also are time insensitive). Latency is a phenomenon not only of mobile networks but also an outcome of all the networks, terminals, and devices through which transmissions may pass, plus the bottlenecks (and, therefore, delays) they may encounter.

Delay can cause protocol time-outs, retransmissions, and disruptions in data circuits and can inhibit voice transmissions. All types of transmission equipment, such as multiplexers, packet assemblers, microwave radios, and digital access cross-connects (DACSs), add small amounts of delay as a result of their internal buffering, and satellite links add significant delay to a signal. ATM switches introduce up to a 2.0 ms delay, and there is a concern about the number of times ATM compression can be used in a tandem link before the overall delay objective is exceeded.

For example, wireless CDMA networks are very sensitive to delays, and vendors recommend that backhaul delay between the cell site and the BSC be below certain limits. For example, the end-to-end delay allowable from the user equipment to a UMTS (WCDMA) radio network controller (RNC) is 7.0 ms. Values of around 12 ms are used in the cdmaOne and CDMA2000 networks.

In addition, utility companies using microwave systems to carry their SCADA signals require a very low delay. That is one reason why electrical utility companies still use old-fashioned analog microwave systems—they introduce smaller delays than digital microwave systems.

The main processing delay in a microwave radio is forward error correction (FEC) buffering, which decreases with an increase in the capacity of the radio link. Total latency of the microwave link is a combination of the radio, free-space, and multiplexer delays. Some typical delays in microwave radio hardware for a point-to-point connection (entire hop) without considering the free-space transit time are shown in the Table 1.6. Microwave radio manufacturers should be able to supply these numbers for their equipment.

TABLE 1.6   Typical Delay Values for
the Microwave Hop

Capacity (Mbps)	Delay (µsec)
2×2	100
4×2	75
8×2	50
16×2	40

1.7 References

1. Lehpamer, H., *Transmission Systems Design Handbook for Wireless Networks*, Norwood, MA: Artech House, 2002.
2. Larus Corporation, *Transmission Engineering Tutorials*, 1996.
3. Flanagan W. A., *The Guide to T1 Networking*, 4th ed., Telecom Library Inc., 1990.
4. ITU-T Recommendation G.702, *Digital Hierarchy Bit Rates*.
5. ITU-T G.803, "Architecture of transport networks based on the synchronous digital hierarchy (SDH)," 03/2000.
6. Goralski, W., *SONET*, 2nd ed., New York: McGraw-Hill, 2000.
7. Zhou, D., "Survivability in Optical Networks," *IEEE Network,* Nov./Dec. 2000.
8. Lewin, B., "SONET Equipment Availability Requirements," *IEEE,* 1989.
9. Malis, A. G., "Reconstructing Transmission Networks Using ATM and DWDM," *IEEE Communications*, Vol. 37, No. 6, June 1999.
10. Bates, J., *Optimizing Voice in ATM/IP Mobile Networks,* New York: McGraw-Hill, 2002,
11. Larus Corporation, *Digital Network Timing and Synchronization*, 1997.

2

Basics of Microwave Communications

2.1 Radio Frequency Spectrum

Radio waves and microwaves are forms of electromagnetic energy we can collectively describe with the term *radio frequency* or *RF*. RF emissions and associated phenomena can be discussed in terms of *energy, radiation,* or *fields*. We can define radiation as the propagation of energy through space in the form of waves or particles. Electromagnetic *radiation* can best be described as waves of electric and magnetic energy moving together (i.e., radiating) through space as illustrated in Fig. 2.1. These waves are generated by the movement of electrical charges such as in a conductive metal object or antenna. For example, the alternating movement of charge (i.e., the *current*) in an antenna used by a radio or television broadcast station or in a cellular base station antenna generates electromagnetic waves. These waves that radiate away from the "transmit" antenna and are then intercepted by a "receive" antenna such as a rooftop TV antenna, car radio antenna, or

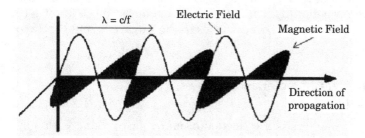

Figure 2.1 Electromagnetic wave.

an antenna integrated into a hand-held device such as a cellular phone.

The term *electromagnetic field* is used to indicate the presence of electromagnetic energy at a given location. The RF field can be described in terms of the electric and/or magnetic field strength at that location. Like any wave-related phenomenon, electromagnetic energy can be characterized by a wavelength and a frequency. The wavelength (λ) is the distance covered by one complete electromagnetic wave cycle, while the frequency is the number of electromagnetic waves passing a given point in one second. Electromagnetic waves travel through space at the speed of light, and the wavelength and frequency of an electromagnetic wave are inversely related by a simple mathematical formula connecting wavelength, speed of light (c), and frequency.

$$\lambda = \frac{c}{f}$$

The wavelength (λ) in centimeters (1 in = 2.54 cm) for a microwave frequency can be simplified as follows:

$$\lambda = \frac{30}{f}$$

where f = frequency in gigahertz (GHz)

Since the speed of light in a given medium or vacuum does not change, high-frequency electromagnetic waves have short wavelengths, and low-frequency waves have long wavelengths. The electromagnetic *spectrum* includes all the various forms of electromagnetic energy from extremely low-frequency (ELF) energy, with very long wavelengths, to X-rays and gamma rays, which have very high frequencies and correspondingly short wavelengths. In between these extremes are radio waves, microwaves, infrared radiation, visible light, and ultraviolet radiation, in that order. The RF part of the electromagnetic spectrum is generally defined as that part of the spectrum where electromagnetic waves have frequencies in the range of about 3 kHz to 300 GHz.

2.2 Structure and Characteristics of the Earth's Atmosphere

The Earth's atmosphere is a collection of many gases along with suspended particles of liquid and solids. Excluding variable components such as water vapor, ozone, sulfur dioxide, and dust, the gases of nitrogen and oxygen occupy about 99 percent of the volume, with argon and carbon dioxide being the next two most abundant gases. From the Earth's surface to an altitude of approximately 80 km, mechanical mixing of the atmosphere by heat-driven air currents evenly distributes the components of the atmosphere. At about 80 km, the mixing decreases to the point at which the gases tend to stratify in accordance with their weights.

The lower, well mixed portion of the atmosphere is called the *homosphere*, and the higher, stratified portion is called the *heterosphere*. The bottom portion of the homosphere is called the *troposphere*. The troposphere extends from the Earth's surface to an altitude of 8 to 10 km at polar latitudes, 10 to 12 km at middle latitudes, and up to 18 km at the equator. It is characterized by a temperature decrease with height. The point at which the temperature ceases to decrease with height is known as the *tropopause*. The average vertical temperature gradient of the troposphere varies between 6 and 7°C/km.

The concentrations of gas components of the troposphere vary little with height, except for water vapor. The water vapor content of the troposphere comes from evaporation of water from oceans, lakes, rivers, and other water reservoirs. Differential heating of land and ocean surfaces produces vertical and horizontal wind circulation that distributes the water vapor throughout the troposphere. The water vapor content of the troposphere rapidly decreases with height. At an altitude of 1.5 km, the water vapor content is approximately half of the surface content; at the tropopause, the content is only a few thousandths of what it is at the surface.

In 1925, the International Commission for Aeronavigation defined the international standard atmosphere. This is a hypothetical atmosphere having an arbitrarily selected set of pressure and temperature characteristics that reflect an average condition of the real atmosphere.

2.3 Radio Propagation

2.3.1 Microwave and Millimeter Waves

Microwave transmission is a very attractive transmission alternative for applications ranging from the coverage of the rural, sparsely populated areas of developing countries that have ineffective or minimal infrastructures to the well developed industrial countries that require rapid expansion of their telecommunications networks. Most of the commercially used terrestrial microwave point-to-point (radio-relay) systems use frequencies from approximately 2 to 60 GHz with maximum hop lengths of around 200 km (125 mi). According to the IEEE, electromagnetic waves between 30 and 300 GHz are called *millimeter waves (MMW)* instead of *microwaves (MW)* because the wavelengths for these frequencies are about 1 to 10 mm. Millimeter-wave propagation has its own peculiarities. The millimeter-wave spectrum at 30 to 300 GHz is of increasing interest to service providers and systems designers because of the wide bandwidths available for carrying communications at this frequency range. Such wide bandwidths are valuable in supporting applications such as high-speed data transmission and video distribution.

Planning for millimeter-wave spectrum use must take into account the propagation characteristics of radio signals at this frequency range. While signals in lower frequency bands can propagate for many miles and penetrate more easily through buildings, millimeter-wave signals can travel only a few miles or less and do not penetrate solid materials very well. However, these characteristics of millimeter-wave propagation are not necessarily disadvantageous. Millimeter waves can permit more densely packed communications links, thus providing very efficient spectrum utilization, and they can increase security of communication transmissions. The radio frequency propagation mechanisms include diffraction, refraction, reflection, and scattering (see Fig. 2.2).

2.3.2 Line-of-Sight Considerations

Microwave point-to-point communications operate in a propagation mechanism called *visibility*, so named for its similarity to light propa-

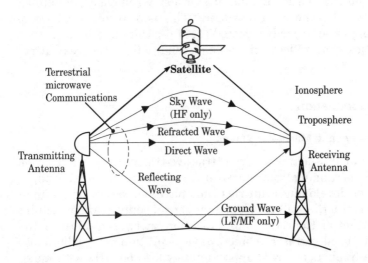

Figure 2.2 Radio wave propagation.

gation. Microwave radio communications require a clear path between parabolic antennas, commonly known as a line-of-sight (LOS) condition. LOS exists when there is a direct path between two separate points and no obstructions (e.g., buildings, trees, hills, or mountains) between them. Partially obstructed paths can also be examined by including the grazing or diffraction loss from the obstruction in the path calculations. Note that unobstructed paths are always preferred, and the outage times and percent reliability computed for obstructed paths may not be reliable.

There is a critical difference between optical LOS (also known as *visual LOS*) and *radio LOS* (or radiovisibility). Visual LOS considers only optical visibility (as seen by the human eye or aided by binoculars) between the ends of the path (see Fig. 2.3). Radio LOS takes into account the concept of Fresnel ellipsoids and their clearance criteria. Under the normal atmospheric conditions, the radio horizon is around 30 percent beyond the optical horizon.

The early nineteenth century French physicist, Augustin Fresnel, made an important observation about the behavior of light. Fresnel noted that a ray of light passing near a solid object is subject to diffraction, or bending. This diffraction causes the intensity of the original light beam to increase or decrease, depending on how near the object is to the beam. This characteristic of electromagnetic radiation is known as the *Fresnel effect,* and light and radio waves are subject to the same laws of physics. If an object such as a mountain ridge or building is close to the radio signal path, it can affect the quality and strength of the signal. Diffraction of the radio waves by such objects can affect the strength of the received signal. This happens even though the obstacle does not directly obscure the direct visual path. This area, known as the *Fresnel zone,* must be clear of all obstructions.

It is possible, for example, to have a microwave path with optical LOS but without radio LOS (according to a certain adopted clearance criterion for that particular path). Radio LOS is more stringent than

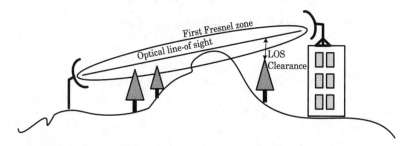

Figure 2.3 Optical and radio line of sight.

optical LOS, since radiovisibility is always considered using the concept of the first Fresnel zone radius along the path. Sometimes, the extra clearance is used when we want to include possible future growth of buildings or trees near the obstacle. Clearance criteria must be followed along the entire microwave path. In the vicinities of antennas, these far-field criteria are substituted by special near-field clearance criteria.

An important objective in planning terrestrial microwave links is to design the radio path in such a way that losses of visibility are extremely rare events. This involves an accurate knowledge of the terrain profile between the terminals and changes in propagation caused by meteorological variations. Sufficient clearance should be guaranteed for the lowest ray to be expected over the path. This can be attained by proper choice of the antenna heights which, however, should not be greater than actually needed—both for economic reasons and because this could increase the risk of fading and signal distortions due to reflections.

2.3.3 Earth Radius and k-Factor

In free space, an electromagnetic wave will travel in a straight line, because the index of refraction is the same everywhere. Within the Earth's atmosphere, however, the velocity of the wave is less than in free space, and the index of refraction normally decreases with increasing altitude. Therefore, the propagating wave will be bent upward or downward from a straight line.

Refraction in the atmosphere is described by its *index of refraction,* which is dependent on the humidity, temperature, and pressure of the atmosphere, all of which are a function of height.[1] Since the barometric pressure and water vapor content of the atmosphere decrease rapidly with height, while the temperature decreases slowly with height, the index of refraction (and therefore refractivity) normally decreases with increasing altitude. As a tool in examining refractive gradients and their effect on propagation, a modified refractivity, defined as

$$M = N + 0.157h \text{ (for } h = \text{altitude in meters)}$$

and

$$M = N + 0.048h \text{ (for } h = \text{altitude in feet)}$$

is often used in place of the refractivity N.

The refractivity distribution within the atmosphere is a nearly exponential function of height (Bean and Dutton, 1968). However, the exponential decrease of N with height close to the Earth's surface (within 1 km) is sufficiently regular to allow an approximation of the exponential function by a linear function, a linear function that is assumed by the effective Earth's radius model. This linear function is known as a *standard gradient* and is characterized by a decrease of 39 *N*-units per kilometer or an increase of 118 *M*-units per kilometer. A standard gradient will cause traveling electromagnetic waves to bend downward from a straight line. Gradients that cause effects similar to a standard gradient but vary between 0 and −79 *N*-units per km or between 79 and 157 *M*-units per km are known as *normal gradients*. Table 2.1 shows normal and anomalous conditions of propagation discussed in more details later in this chapter.

TABLE 2.1   Refractive Gradients and Conditions

	N-gradient	M-gradient
Trapping	< −157 N/km	< 0 M/km
	< 48 N/kft	< 0 M/kft
Superrefractive	−157 to −79 N/km	0 to 79 M/km
	−48 to −24 N/kft	0 to 24 M/kft
Normal	−79 to 0 N/km	79 to 157 M/km
	−24 to 0 N/kft	24 to 48 N/kft
Subrefractive	> 0 N/km	> 157 N/km
	> 0 N/kft	> 48 N/kft

To facilitate path profiling at the time computers were not available, radio transmission engineers introduced the Earth-radius factor k to compensate for the refraction in the atmosphere. Applying appropriate k-values to the true-Earth radius, an equivalent-Earth radius is geometrically obtained and, consequently, straight rays can be drawn (see Fig. 2.4). The image shows an equivalent Earth with different ray beam curvatures for different values of the Earth-radius factor.

The effective Earth radius factor, k, is defined as the factor that is multiplied by the actual Earth radius, a, to give the effective Earth radius a_e. The mean Earth radius is in average 6,371 km.

Due to the Earth's curvature and refraction of radio signals by objects, each site must meet a minimum elevation with respect to antenna height. The effects of refraction are significant within an area around the direct path known as the Fresnel zone. The maximum ef-

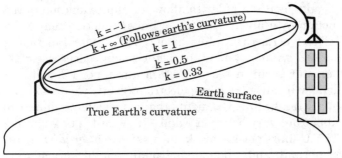

Effective Earth's Radius = k × True Earth's Radius

True Earth's Radius = 6,371 km

k = 1 used for establishing LOS in most parts of the world
k = 4/3 standard atmosphere with normally refracted path
k < 1 "Earth's bulge with path obstruction" is rare but possible
    in some climates
−1 < k < ∞ rare occurrence of entrapment layers in some areas

Figure 2.4 Variations of the ray curvature as a function of k.

fects caused by the Earth's curvature and Fresnel zone occur at the
midpoint of the link. It is usually adequate to use less than the full
depth of the Fresnel zone to calculate clearance, and 60 percent of the
first Fresnel zone is the generally accepted area that must be kept
clear of obstructions.

The Earth-radius factor k for a certain area can be calculated based
on refractivity gradient found from the local charts. For standard refractivity conditions, $k = 1.33 = 4/3$, and this value should be used
whenever local value is not provided. It is important to keep in mind
that lower k-values will lower the LOS; in other words, demand higher
antenna heights. Some typical values of k for the U.S.A. are shown in
the Table 2.2.

TABLE 2.2   Typical Values of k in the U.S.A.

	Summer	Winter
Dry mountains (above 1500 m)	1.20	1.20
Mountains (to 1500 m)	1.25	1.25
Midwest and Northeast	1.50	1.30
South and West Coast	1.55	1.35
Southern Coast	1.60	1.50

2.3.4   Standard Propagation Mechanisms

Standard propagation mechanisms are those mechanisms and processes that occur in the presence of a standard atmosphere. These propagation mechanisms are free-space propagation, reflection, diffraction, scattering, and tropospheric scatter.

2.3.4.1   Propagation in a homogeneous atmosphere. The simplest case of electromagnetic wave propagation is the transmission of a wave between a transmitter and a receiver in a homogeneous atmosphere (sometimes also called *free space*). It is defined as a region whose properties are isotropic, homogeneous, and loss-free; i.e., away from the influences of the Earth's atmosphere. In free space, the electromagnetic wavefront spreads uniformly in all directions from the transmitter. If a particular point on a wavefront were followed over time, the collection of point positions would define a ray. The ray would coincide with a straight line from the transmitter to the receiver.

2.3.4.2   Reflection. Reflection occurs when an electromagnetic wave strikes a nearly smooth, large surface, such as a water surface, and a portion of the energy is reflected from the surface and continues propagating along a path that defines an angle with the surface equal to that of the incident ray. Obstruction dimensions are very large compared to the signal wavelength.[2] The strength of the reflected wave is determined by the reflection coefficient, a value that depends on the frequency and polarization of radiation, the angle of incidence, and the roughness of the reflecting surface. For shallow incidence angles and smooth seas, typical values of the reflection coefficient are near unity (i.e., the reflected wave is almost as strong as the incidence wave). Reflection rays from different surfaces may interfere constructively or destructively at a receiver (multipath propagation).

2.3.4.3   Diffraction. Diffraction occurs when an impenetrable body obstructs the radio path between the transmitter and receiver. Energy tends to follow along the curved surface of an object. Based on Huygens' principle, secondary waves form behind the obstructing body, and the study is usually mainly based on physical optics. The ability of the electromagnetic wave to propagate beyond the horizon by diffraction is highly dependent on frequency; the lower the frequency, the more the wave is diffracted.

When the clearance of the radio path over the underlying terrain becomes small, diffraction phenomena take place, and they reduce the received signal strength. To determine how close the radio path can approach an obstacle before diffraction losses begin to occur, we can use the concept of the first Fresnel zone.

2.3.4.4 Scattering. Scattering occurs when the radio channel contains objects whose dimensions are approximately the same as or smaller than the propagating wavelength. Scattering, which follows the same physical principles as diffraction, causes energy to be radiated in many different directions. As frequencies increase, the wavelengths become shorter, and the reflective surface appears rougher, thus resulting in more diffused reflections as opposed to specular reflections.

At ranges far beyond the horizon, the propagation loss is dominated by troposcatter. Propagation in the troposcatter region is the result of scattering by small imperfections within the atmosphere's refractive structure.

2.3.5 Anomalous Propagation Mechanisms

Anomalous meteorological conditions may occur that considerably change the standard propagation described above. A deviation from the normal atmospheric refractivity leads to conditions of anomalous propagation such as subrefraction, superrefraction, and trapping (ducting). It is hard to predict, and almost impossible to completely avoid, the effects of anomalous propagation.

2.3.5.1 Subrefraction. If the motions of the atmosphere produce a situation in which the temperature and humidity distribution creates an increasing value of N with height, the wave path would actually bend upward, and the energy would travel away from the Earth. This is termed *subrefraction*. Although this situation rarely occurs in nature, it still must be considered when assessing electromagnetic systems' performance.

Subrefractive layers may be found at the Earth's surface or aloft. In areas where the surface temperature is greater than 30°C, and relative humidity is less than 40 percent (i.e., large desert and steppe regions), solar heating will produce a very nearly homogeneous surface layer, often several hundreds of meters thick. Since this layer is unstable, the resultant convective processes tend to concentrate any available moisture near the top of the layer, which in turn creates a positive N gradient. This layer may retain its subrefractive nature into the early evening hours, especially if a radiation inversion develops, trapping the water vapor between two stable layers.

For areas with surface temperatures between 10 and 30°C and relative humidity above 60 percent (e.g., the western Mediterranean, Red Sea, Indonesian Southwest Pacific), surface-based subrefractive layers may develop during the night and early morning hours, caused by advection of warm, moist air over a relatively cooler and drier surface. While the N gradient is generally quite intense, the layer is often not very thick.

2.3.5.2 Superrefraction. Superrefractive conditions are largely associated with temperature and humidity variations near the Earth's surface. The effects of a superrefractive layer on a microwave system is directly related to its height above the Earth's surface. If the troposphere's temperature increases with height (temperature inversion) and/or the water vapor content decreases rapidly with height, the refractivity gradient will decrease from the standard. The propagating wave will be bent downward from a straight line more than normal. As the refractivity gradient continues to decrease, the radius of curvature for the wave path will approach the radius of curvature for the Earth. The refractivity gradient for which the two radii of curvature are equal is referred to as the *critical gradient*. At the critical gradient, the wave will propagate at a fixed height above the ground and will travel parallel to the Earth's surface. Refraction between the normal and critical gradients is known as *superrefraction*.

2.3.5.3 Ducting. Should the refractivity gradient decrease beyond the critical gradient, the radius of curvature for the wave will become smaller than that of the Earth. The wave will either strike the Earth and undergo surface reflection or enter a region of standard refraction and be refracted back upward, only to reenter the area of refractivity gradient that causes downward refraction.[3] This refractive condition is called *trapping* (or *blackout fading*), because the wave is confined to a narrow region of the troposphere. The common term for this confinement region is a *tropospheric duct* or a *tropospheric waveguide*. Trapping is an extension of superrefraction, because the meteorological conditions for both are the same. In a discussion of atmospheric ducting conditions on electromagnetic wave propagation, the usual concern is propagation beyond the normal horizon.

To propagate energy within a duct, the angle the electromagnetic system's energy makes with the duct must be small—usually less than 1°. Thicker ducts, in general, can support trapping for lower frequencies. The vertical distribution of refractivity for a given situation must be considered as well as the geometrical relationship of transmitter and receiver to the duct so as to assess the duct's effect at any particular frequency.

Several meteorological conditions could lead to the creation of ducts. If these conditions cause a trapping layer to occur such that the base of the resultant duct is at the Earth's surface, a *surface duct* is formed. Surface-based ducts occur when the air aloft is exceptionally warm and dry compared with the air at the Earth's surface. Several meteorological conditions may lead to the formation of surface-based ducts. For example, over the ocean and near land masses, warm, dry continental air may be blown over the cooler water surface such as the

Santa Ana of southern California, the Sirocco of the southern Mediterranean, and the Shamal of the Persian gulf. This advection will lead to a temperature inversion at the surface. In addition, moisture is added to the air by evaporation, producing a moisture gradient to strengthen the trapping gradient. This type of meteorological condition routinely leads to a surface duct created by a surface-based trapping condition. Surface-based ducts tend to be on the leeward side of land masses and may occur both during the day or at night. In addition, surface-based ducts may extend over the ocean for several hundred kilometers and may be very persistent (lasting for days).

Elevated ducts may vary from a few hundred meters above the surface at the eastern part of the tropical oceans to several thousand meters at the western part. For example, along the southern California coast, elevated ducts occur an average of 40 percent of the time, with an average top elevation of 600 m. Along the coast of Japan, elevated ducts occur an average of 10 percent of the time, with an average top elevation of 1,500 m. It should be noted that the meteorological conditions necessary for a surface-based duct are the same as those for an elevated duct.

Evaporation ducts exist over the ocean, to some degree, almost all of the time. The duct height varies from a meter or two in northern latitudes during winter nights to as much as 40 m in equatorial latitudes during summer days. On a world average, the evaporation duct height is approximately 13 m. The duct strength is also a function of wind velocity. For unstable atmospheric conditions, stronger winds generally result in stronger signal strengths (or less propagation loss) than do weaker winds. Since the evaporation duct is much weaker than the surface-based duct, its ability to trap energy is highly dependent on frequency. Generally, the evaporation duct is only strong enough to affect electromagnetic systems above 3 GHz.

It is well known fact that the equatorial regions are most vulnerable to ducts. In temperate climates, the probability of formation of ducts is lower. The ducting probability follows seasonal variations. Conventional techniques used to combat other types of fading, such as increased margins or diversity techniques, have little or no influence on blackout fading.

2.3.6 About Propagation Conditions

Propagation conditions vary from month to month and from year to year, and the probability of occurrence of these conditions may vary by as much as several orders of magnitude. From propagation data, it was concluded that, for a well designed path that is not subject to diffraction fading or surface reflections, multipath propagation is the dominant factor in fading below 10 GHz. Above this frequency, the effects of precipitation tend increasingly to determine the permissible path length through the system availability objectives. The necessary reduction in path length with increase in frequency reduces the severity of multipath fading. These two principal causes of fading are normally mutually exclusive. Given the split between availability and error performance objectives, precipitation effects contribute mainly to unavailability and multipath propagation mainly to error performance. Another influence of precipitation (e.g., backscatter from rain) may influence the choice of radio frequency channel arrangements.

Propagation effects due to various forms of precipitation tend not to be frequency dispersive, while multipath propagation caused by tropospheric layers can be, and this may cause severe distortion of information-bearing signals. The rapid development of digital communication systems has required an improved understanding of these effects and the means to overcome them.

Two countermeasures to propagation distortion are commonly used: diversity techniques and adaptive channel equalizers, which attempt to combat attenuation and distortion caused by the propagation medium. The effectiveness of a fading countermeasure is usually expressed in terms of an improvement factor.

2.4 Digital Microwave Point-to-Point Systems

2.4.1 Microwave Radio Basics

A significant proportion of business premises lack broadband connectivity, so *wireless* access provides the perfect medium for connecting new customers. Even if an operator chooses to use unlicensed or point-to-multipoint wireless technologies to connect customers, high-capacity microwave provides the ideal solution for backhaul of customer traffic from access hubs to the nearest fiber point of presence (POP).

Wireless operators and *wireless networks* often face aggressive schedules to provide service for customers and to generate immediate revenue. To turn up their networks, they need to connect their cell sites to switching stations, and they have chosen microwave because of its reliability, speed of deployment, and cost benefits over fiber or leased-line alternatives. Microwave radio is deployed in the emerging 2.5 and 3G mobile infrastructures to support increased data usage and greater numbers of cell sites needed to support a new generation of mobile services.

Companies now have *private networks* with high-speed LAN/WAN network requirements and need to connect parts of their businesses in the same campus, city, or country. Utility companies use microwave systems to connect remote sites, control gas pipelines, and electric powerline networks and to tie together a corporation's several operating locations.

Public transport organizations, railroads, and other public utilities are major users of microwave systems. These companies also use microwave systems to carry control and monitoring information *(SCADA systems)* to and from power substations, pumping stations, and switching stations.

Many countries, especially developing countries, need to upgrade their telephone infrastructure systems to the latest digital technology. Microwave has traditionally provided developing nations with the tools of establishing telecommunications quickly over often undeveloped and impractical terrain such as deserts, jungle, and frozen terrain where laying cable would be a very difficult and/or expensive undertaking.

Governments in many countries use fixed services to provide a number of defense and nondefense functions in locations and circumstances where commercial communication services do not satisfy requirements. They use microwave point-to-point services for control and monitoring of many wide-area systems (e.g., air traffic control) for connecting government mobile radio sites, for tactical communications, and for communications within test and training ranges.[4]

Microwave point-to-point communication can be achieved by a single connection—for example, a microwave link between two stations located at specified fixed points or multiple cascaded links made by a number of intermediate repeaters with or without partial payload drop-insert. The transmitted information can be voice, data, or video as long as it is in a digital format. A typical digital microwave radio consists of three basic components.

- A digital modem for interfacing with digital terminal equipment

- A radio frequency (RF) unit for converting a carrier signal from the modem to a microwave signal

- An antenna used to transmit and receive the signal

The combination of these three components is called a *radio terminal.* Two terminals are required to establish a microwave communications link, commonly referred to as a *microwave hop* or *microwave link* (Fig. 2.5).

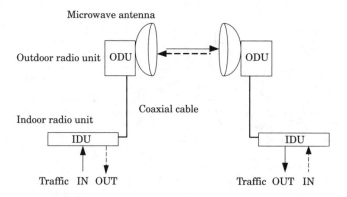

Figure 2.5 Typical point-to-point MW system.

We can feed the data and voice traffic into the radio using an electrical or optical line. In the radio, the digital signals are coded into analog signals and converted to microwaves (with a typical wavelength of a few centimeters). The microwaves are sent using a highly directive parabolic shaped antenna. At the other end, the signals are received and restored to the digital format.

The plesiochronous digital hierarchy (PDH) microwave radio provides a transmission medium for digital traffic of standard capacities typically ranging from 1.544 Mbps (1T1) to 45 Mbps (1DS3) in North America and from 2.048 Mbps (1E1) to 34 Mbps (16E1) based on ITU standards. Sometimes, high-capacity backbones in North America are built for 3DS3 and require at least 28 MHz of bandwidth. These radio links may be established between any two points within the line of sight and, depending on the frequency, geographical region, and rain statistics, the typical link distance can be up to 25 mi (40 km). For the longer microwave link hops, additional measures have to be taken to ensure required reliability of the system (e.g., space and/or frequency diversity). In today's wireless networks, these PDH microwave systems are used for the low- to medium-capacity links (e.g., RBS-BSC backhaul connectivity in wireless networks).

Microwaves, which are only centimeters (or inches) in length, are small relative to the surroundings and hence do not have the bending property. Therefore, to establish a radio link, it is important to have radio LOS between the two radio position sites. One or more radio paths connected in tandem form a *microwave system*. The radio stations between two terminal stations are called *repeater stations* (active or passive) (see Fig. 2.6).

Repeaters can be *nonregenerative* when the signal is only filtered and amplified, with or without down and up conversions (e.g., in some

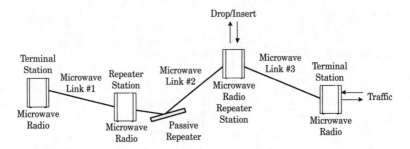

Figure 2.6 Radio link with repeater.

analog FDM systems) or *regenerative* when, in digital applications, the signal is demodulated and remodulated before transmission to the next radio hop.

Passive repeaters implemented without any active radio components (e.g., two directional antennas connected back to back, a reflector, and so on) are also utilized.

Drop-insert is a functionality provided in analog and digital repeaters, where only radio-system specific control and service channels, and possibly part of the payload, is made available for local traffic and system management and maintenance.

2.4.2 Fresnel Zones and Clearance Rules

The most common use of Fresnel zone information on a profile plot is to check for obstructions that penetrate the zone. As emphasized earlier, although line of sight is important, it may not always be adequate. Even though the path has clear line of sight, if obstructions (such as terrain, vegetation, buildings, and others) penetrate the Fresnel zone, there will be signal attenuation. The higher the frequency, the narrower the Fresnel zone and, consequently, the more vulnerability to non-LOS effects (object attenuation). Fresnel zones are specified employing an ordinal number that corresponds to the number of half-wavelength multiples that represents the difference in radio wave propagation path from the direct path. The first Fresnel zone is therefore an ellipsoid whose surface corresponds to one half-wavelength path difference and represents the smallest volume of all the other Fresnel zones. The radius of the first Fresnel zone (Fig. 2.7) is the parameter currently employed to establish appropriate clearance.

The Fresnel zone is computed along the path, usually for the distance of each of the terrain points, so the resolution of the computed and plotted Fresnel zone is comparable to the terrain data. Typically, the first Fresnel zone ($N = 1$) is used to determine obstruction loss,

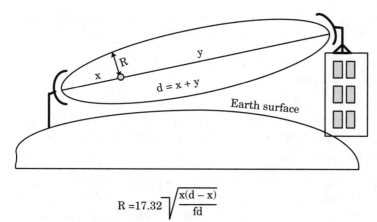

$$R = 17.32 \sqrt{\frac{x(d - x)}{fd}}$$

d = distance between antennas (hop height in kilometers)
R = first Fresnel zone radius in meters
f = frequency in Gigahertz

Figure 2.7 Radius of the first Fresnel zone.

and anytime the path clearance between the terrain and the line-of-sight path is less than $0.5F_1$ (half of the first Fresnel zone radius), some *knife-edge diffraction* loss will occur. The amount of loss depends on the amount of penetration. Profiles are often drawn with the first Fresnel zone and a ratio of 0.5 to provide a quick visual inspection of possible problems caused by obstructions penetrating that zone. Some engineers plot a ratio of 0.6 of the first Fresnel zone to add a bit of headroom for the path design.

The refractive properties of the atmosphere are not constant. The variations of the *index of refraction* in the atmosphere (expressed by the Earth-radius factor k) may force terrain irregularities to totally or partially intercept the Fresnel zone. *Clearance* can be described as any criterion to ensure sufficient antenna heights so that, in the worst case of refraction (for which k is minimum), the receiver antenna is not placed in the diffraction region.

Diffraction theory indicates that the direct path between the transmitter and the receiver needs a clearance above ground of at least 60 percent of the radius of the first Fresnel zone to achieve free-space propagation conditions. Clearance values have to fit the local climate. Clearance can be considered by applying climate-dependent *clearance criteria* or by properly handling diffraction-diffraction fading (k-type fading). Recently, with more information on this mechanism and the statistics of k_e that are required to make statistical predictions, some administrations are installing antennas at heights that will produce some small known outage.

To summarize all this, we can say that, for normal propagation conditions, the following clearance criteria have to be satisfied:

- Clearance of 60 percent or greater at the minimum k suggested for the certain path, and
- Clearance of 100 percent or greater at $k = 4/3$.
- In case of space diversity, the antenna can have a 60 percent clearance at $k = 4/3$ plus allowance for the tree growth, buildings, and so forth—usually 3 m (10 ft).

Another important use of Fresnel zone information is to check microwave paths for possible reflection points. For even-numbered Fresnel zones ($N = 2, 4, ...$), the difference between the direct path and the indirect path defined by the Fresnel zone radius is a multiple of one half-wavelength. If the geometry of the path is such that an even-numbered Fresnel zone happens to be tangential to a good reflecting surface (e.g., a lake, highway, or smooth desert area, depending on what wavelength is involved), signal cancellation will occur as a result of interference between the direct and indirect (reflected) signal paths. It is possible to set the Fresnel zone to even-numbered values when plotting a profile to see if any potential areas of destructive signal reflection are present on the path.

2.4.3   Near and Far Fields

The terms *far field* and *near field* describe the fields around an antenna or, more generally, any electromagnetic radiation source. The names imply that two regions, with a boundary between them, exist around an antenna. Actually, as many as three regions and two boundaries exist. These boundaries are not fixed in space, and they move closer to or farther from an antenna, depending on both the radiation frequency and the amount of phase error an application can tolerate. In the literature, these regions have different names and can be defined in slightly different ways. Usually, two- and three-region models are used. The near field may be thought of as the transition point where the laws of optics must be replaced by Maxwell's equations of electromagnetism. In the three-region model, near field, far field, and the transition zone are defined as follows.

The *near field,* also called the *"reactive near field,"* is the region that is closest to the antenna and for which the reactive field dominates over the radiative fields. In the reactive near-field region, fields vary as $1/r^3$ (power varies as $1/r^6$). For antennas that are large in terms of wavelength, the near-field region consists of the reactive field extending to the certain distance followed by a radiating near field.

The *transition zone* or *"radiating near field"* is the region between the reactive near field and the far-field regions and is the region in which the radiation fields dominate and where the angular field distribution depends on distance from the antenna. In the radiating near-field region, fields vary with $1/r^2$.

In the radiating near field, the field strength does not necessarily decrease steadily with distance away from the antenna but may exhibit an oscillatory character. Therefore, it is difficult to predict the antenna gain in that region.

The *far field,* or Rayleigh distance, is the region where the radiation pattern is independent of distance. In this area, fields vary with $1/r$.

$$Far\ field = 2D^2/\lambda\ to\ infinity$$

where D = the largest dimension of the antenna

At this point, results of measuring real antenna gain would have less than 1 percent error as compared to the real gain. Although formulae for the near-field and transition-zone boundary vary in the literature, they all agree on the far-field boundary.

Equations contain terms in $1/r$, $1/r^2$, and $1/r^3$. In the near field, the $1/r^3$ terms dominate the equations. As the distance increases, the $1/r^3$ and $1/r^2$ terms attenuate rapidly and, as a result, the $1/r$ term dominates in the far field. It is also important to notice that far-field expressions are valid if $D \gg \lambda$, which is always a case in microwave systems.

Engineers perform link engineering, including Fresnel clearances and path profiles, based on the assumption that microwave antennas are in the far-field region. For example, a microwave system operating in an 8 GHz band and having 6-ft dishes will have a far field beginning at 213 m (634 ft) from the antenna using general formula for the far-field boundary.

2.4.4 Link Budget

A microwave engineer starts the microwave link design by doing a link budget analysis. The link budget is a calculation involving the gain and loss factors associated with the antennas, transmitters, transmission lines, and propagation environment, used to determine the maximum distance at which a transmitter and receiver can successfully operate.

The *receiver sensitivity threshold* is the signal level at which the radio runs continuous errors at a specified bit rate. Specifications are listed for the 10^{-3} bit error rate (PDH radios).

System gain (in decibels) is defined as the difference between the transmitter output power and the receiver threshold. Lowering the system gain will reduce the fade margin. System gain can be used to reduce antenna sizes or improve the path availability. A given radio system has a system gain that depends on the design of the radio and the modulation used. For example, 99.999 percent system availability (5 min outage per year) will degrade to 99.980 percent (2 hr outage per year) if the modulation is changed from 16 QAM to 128 QAM without recovering the system gain reduction and all other conditions remaining unchanged.

The gains from the antenna at each end are added to this gain, with larger antennas providing higher gain. The free-space loss of the radio signal as it travels over the air is then subtracted from the system, and the longer the link, the higher the loss. These calculations result in a "fade margin" for the link (see Fig. 2.8). In most applications, the same duplex radio setup is applied to both stations forming the radio-relay path. Thus, the calculation of the received signal level is independent of direction.

The fade margin is calculated with respect to the receiver threshold level for a given bit-error ratio (BER). The threshold level for BER =

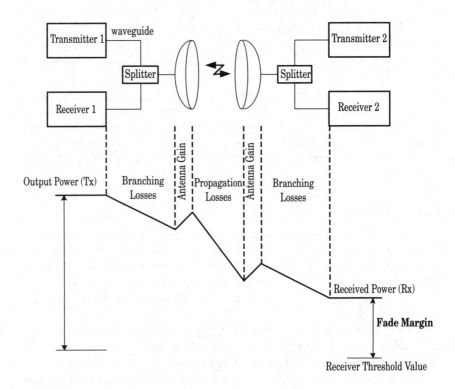

Figure 2.8 Radio path link budget.

10^{-6} for microwave equipment used to be about 3 dB higher than the threshold level for BER = 10^{-3}. Consequently, the fade margin was 3 dB larger for BER = 10^{-6} than for BER = 10^{-3}. For the new generation of microwave radios with power forward *error correction schemes,* this difference is more in the 0.5 to 1.5 dB range.

The radio can handle anything that affects the radio signal within the fade margin. If the margin is exceeded, then the link could go down and therefore become unavailable.

The next step, then, is to analyze rain fade, multipath fading, interference, and other miscellaneous losses that could potentially affect the radio signal.

2.4.5   Microwave Systems for Rapid Deployment

"Licensed" RF transmitters communicate using a specific transmit and receive frequency combination that is assigned to the user (licensee). The frequency assignment is coordinated with other users of the same spectrum in the same geographical area. This process provides full disclosure of the frequency assignment and typically avoids interference from any existing licensee in the area. If licensed radios encounter interference, it is typically resolved with the assistance of the regulatory body.

The terms *license-exempt* (also called *license-free* or *unlicensed*) and *licensed* refer to the radio frequency spectrum rules defined by the U.S. Federal Communications Commission (FCC) or an equivalent national government regulatory body. In the United States, FCC Rules Part 15 governs the license-exempt frequency spectrum, and Rules Part 101 governs the licensed frequency spectrum. Licensed products require regulatory approval before deployment, whereas license-exempt products can be deployed without any regulatory approval.

A *license-exempt* system can be installed virtually anywhere within a given country without obtaining an operational license from the regulatory authorities. As such, they are good candidates for the *rapid deployment microwave systems.* Of course, such a system must already be certified to operate as license-exempt in that country. Manufacturers desiring license-exempt certification must apply to the FCC or equivalent national authority for the *type approval* to operate the particular product in specific radio frequency bands.

2.4.5.1   Spread-spectrum microwave radios.   Spread-spectrum systems were originally developed by the military to counter attempts to detect, decode, or block signal transmissions. The most important non-

military characteristics of spread spectrum systems are that they facilitate radio communications in a manner that minimizes the potential to cause harmful interference to other services, and they are able to withstand higher levels of interference than other technologies. Hence, spread-spectrum systems have significant potential to share common spectrum with other services. Spread-spectrum devices operate on a license fee-exempt basis, if the technical conditions are met, and type approval is mandatory. Two main types of spread-spectrum system are commercially available: *direct sequence* and *frequency hopping.*

Direct sequence (DS) is where the incoming information is usually digitized (if not already in binary format) and then modulo 2 added to a higher-speed code sequence. The combined information and the code are then used to modulate a radio frequency carrier. Since the high-speed code sequence dominates the modulating function, it is the direct cause of the wide spreading of the transmitted signal.

Frequency hopping (FHSS) is where the carrier is modulated with coded information in a conventional manner, causing a conventional spreading of the radio frequency energy about the carrier frequency. The frequency of the carrier is not fixed but changes rapidly under the direction of a pseudorandom coded sequence. The wide radio frequency bandwidth needed by such a system is not required to support a spreading of the radio frequency energy about the carrier, but rather to accommodate the range of frequencies to which the carrier frequency can hop.

A unique pseudorandom code may be embedded in the signal during the modulation process, enabling a large number of users to occupy the same band, as each receiver will recognize only its own code. This access technique is called *code division multiple access (CDMA).* Typical bandwidths used by a spread spectrum system range from 500 kHz up to 50 MHz, depending on the data throughput required and the bandwidth available. The bandwidth of the transmitted signal is much greater than the bandwidth of the original message, and the bandwidth of the transmitted signal is determined by the message to be transmitted and by an additional signal known as the *spreading code.*

The frequencies requested by license applicants for operation of spread-spectrum devices have been mainly in bands designated for *industrial, scientific, and medical (ISM)* applications at 900 MHz, 2.4 GHz, and (to a lesser extent) at 5.8 GHz. ISM devices generate and internally use radio frequency energy for industrial, scientific, medical, domestic, or similar purposes, but not for communications. Examples of ISM devices are plastic welders, chemical analysis equipment, medical diathermy equipment, wireless microphones, and domestic microwave ovens. These devices radiate incidental emissions that have the potential to cause interference to radio communication equipment. Bands for ISM applications have been specifically designated, via footnotes in the Spectrum Plan (and the ITU Radio Regulations) for the operation of such equipment. In these bands, radio communications services are not protected against any interference caused by ISM equipment.

Spread-spectrum communications microwave systems should be used with caution and only as a temporary solution when rapid deployment is required and/or in rural areas only. The reason is that the 2.4-GHz band is also used with a number of other applications within the unlicensed ISM band, including garage door openers, microwave ovens, Bluetooth systems, Wireless LANs, and so forth. The 2.4-GHz ISM band used to be called the "junk band," because it was already contaminated by oven emissions. Years ago, 2.43 GHz was allocated to the microwave oven, and it was felt that no one would ever want to co-occupy this band. As pressure to allocate more spectrum to communications was felt, the FCC set up rules for unlicensed ISM operation in this band.

Commercial unlicensed spread-spectrum systems typically use the ISM bands worldwide. These are located as shown below:

- 902 to 928 MHz

- 2400 to 2483.5 MHz (microwave ovens are located here)

- 5725 to 5850 MHz

Those in Canada are similar and are based on Industry Canada, Spectrum Management, RSS-210, "Low Power License-Exempt Radiocommunication Devices."[*]

Licensed microwave systems in Canada using the same band are described in RSS139, "Licensed operation from 2400 to 2483.5 MHz."

In part for reasons of safety, the transmitter power output level in the ISM band is limited to 1 W (+30 dBm) maximum input antenna power. For similar reasons, and to minimize interference, *effective isotropic radiated power (EIRP)*, or power radiated by the associated antenna system, is limited to 4 W (+36 dBm) maximum in Canada.

In the U.S. (2.4 GHz band), for every 3 dB of antenna gain over +6 dBi, the input power to the antenna is reduced by 1 dB from +30 dBm. In the 5.8 GHz band, there are no EIRP limits—only a +30 dBm maximum antenna input power.

---

[*]More information about Industry Canada can be found on its Web page: http://strategis.ic.gc.ca/sc_mrksv/spectrum/engdoc/spect1.html).

It is important to remember that some licensed users sometimes operate in the unlicensed bands as well. The unlicensed bands are allocated on a shared basis, and, while there may be no requirement to obtain a license to operate for low-power datacom applications with approved equipment, other licensed users may be allowed to operate with significantly higher power. A specifically important example of this is operation of U.S. government radar equipment in the U.S. U–NII band at 5.725 to 5.825 GHz. These radars often operate at peak power levels of millions of watts, which can cause significant interference problems to other nearby users in this band. Therefore, it is important to look around the site to determine if there are any airports, military bases, or other installations where such radars may be located.

Although the lack of licensing requirements for the ISM band implies ease of installation and minimal capital outlay, it by no means implies a lack of regulatory controls. Most regulations permitting unlicensed spread spectrum penalize the use of directional antennas, and manufacturers using directional antennas are forced to reduce the transmitter output power by the gain of the antenna to comply with an effective radiated power (ERP) limit. From the regulatory perspective, these unlicensed bands come with two major constraints.

- Transmit power limitation of 1 W

- Minimum processing gain of 10 dB for either FHSS or DS

This implies that the desired data capacity per bandwidth (in other words, bandwidth efficiency) may have to be sacrificed to achieve the processing gain, and the total transmit power is not high enough to support multilevel quadrature-amplitude-modulation (QAM) techniques to increase the data rate. The challenge is to overcome these difficulties and still achieve sufficiently high data capacity.

For FHSS systems, IEEE 802.11 defines 79 different hops for the carrier frequency. Using these 79 frequencies, IEEE 802.11 defines 78 hopping sequences (each with 79 hops) grouped in three sets of 26 sequences each. Sequences from same set encounter minimum collisions, and they may be allocated to collocated systems. Theoretically, 26 FHSS systems may be collocated. However, as synchronization among independent systems is forbidden (synchronization would eliminate collisions), the actual number of systems that can be collocated is around 15.

2.4.5.2 Cell on Wheels (COW). Figure 2.9 shows a portable wireless cell-site trailer. During the wireless network launch, operators bring

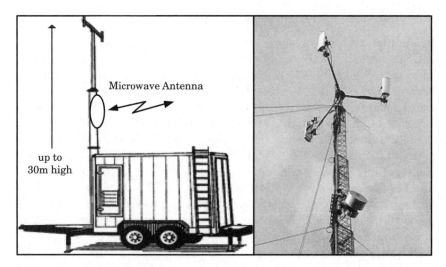

Figure 2.9 Cell-on-wheels (COW).

out these portable towers on a trailer and set them up as a temporary fix until a more permanent site can be found. The cell on wheels can be equipped with up to a 30-m-long hydraulically expandable mast. They are self-sufficient and capable of operating in the field for extended periods of time, and one person can usually set one up in less than an hour.

Natural and man-made disasters can be catastrophic to communications networks. Microwave systems are often used to restore communications when other transmission equipment has been damaged by earthquakes, floods, hurricanes, or other natural disasters; conflicts such as terrorist attack and wars; and other wireless network problems.

COW is usually connected via a leased T1/E1 backhaul circuit with the switch or via microwave link to the nearest working cell-site. In case of microwave connection, licensed or unlicensed microwave radios can be used, but the preference would be to use an unlicensed spread-spectrum microwave link.

The military uses what it calls a *tactical transportable antenna system (TTAS)* for applications in which rapid deployment is a critical concern. The system consists of a 3-m-diameter reflector mounted on a modified military trailer. The system provides ground-level elevation and azimuth controls and is supplied with guy wires, ground anchors, and waveguide termination behind the reflector. A trained team of two people can erect the antenna in less than 20 min. The system is usually used for line-of-sight or over-the-horizon microwave communications in the 5 GHz band.

2.4.6   Over-the-Horizon Microwave Systems

In addition to terrestrial systems operating within the radio horizon (LOS), there is a different type of point-to-point microwave system. These are called *over-the-horizon (OH)* systems (see Fig. 2.10) and are described in ITU-R P.617. For links within the radio horizon, the antennas should be placed to provide direct line-of-sight transmission. Practically, this means that the equipment should be installed on towers or buildings of a certain height that fulfill the line-of-sight requirement and minimize the probability of the fading due to multipath propagation. For over-the-horizon links, a virtual tower height is defined either by the propagation mechanism itself (scattering capability of the troposphere, reflection capability of the meteorite tail) or by the physical placement of the equipment (stratosphere platform, satellite platform).

The *troposphere* is the lowest 10-km part of the atmosphere, and the inhomogenities within the common beam volume of the transmit and receive antennas act as sources that scatter the electromagnetic waves into all directions of space. Thus, components will be generated establishing coupling between transmitter and receiver. The first troposcatter link was put into operation in 1953, and the link allows distances of up to 400 to 500 km to be spanned. Troposcatter microwave links have been applied in locations that difficult to access, such as links connecting oil well islands with the mainland, links within deserts and jungles, and links for military communication (before the

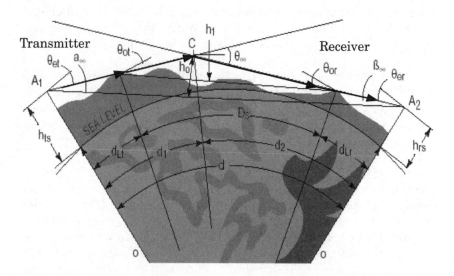

Figure 2.10   Over-the-horizon MW systems.

large-scale use of inexpensive satellite links). These links are operating in the 350 MHz to 6 GHz range within the bands allocated by the ITU. The parameters of the radio channel require extremely high-gain antennas and transmitting powers, plus the application of multiple diversity systems such as frequency and space diversity, and angle and frequency diversity. The capacity of the transmitted information is limited by propagation time differences over the scattering volume.

In addition to the more commonly used over-the-horizon microwave point-to-point systems based on tropospheric scattering, there are other point-to-point systems based on stratosphere platform systems (HAP) and systems utilizing meteorite tails (ITU-R P.843) that cannot be called microwave, since they operate in much lower frequency bands.

2.5 Other Microwave Systems

So far, we have talked about point-to-point microwave systems (often simply called *microwave systems*) that provide wideband communications over the line-of-sight paths. Microwave technology is becoming increasingly useful for accessing individual users and narrow- and medium-bandwidth groups of users, thereby replacing wireline or fitting into the bandwidth gap between wireline and fiber. Expectations are to see more microwave applications for WLL, WLANs, Bluetooth, and other new short-range access technologies. Many of these are non-traditional microwave applications, however, and they are not as easily recognized as the traditional microwave fixed services of the past. The new applications will tend to move the microwave industry closer to the individual user and closer to the mass market. These short-range access markets will be particularly suitable for higher microwave frequency bands where the smaller antenna size, greater frequency reuse, and wider available bandwidths provide major advantages.

It is important to note that 3G technologies (WCDMA, CDMA2000), WLAN, and Bluetooth are not competing technologies but complement each other in creating an overall wireless network and operating in the 2-GHz microwave band.

2.5.1 Point-to-Multipoint Systems

The existing telecom networks were designed for the delivery of voice services. As the demand for Internet traffic has increased, the access portion of the network has been unable to achieve the desired speeds because of limitations in the deployed technology. Copper wires, for example, over typical distances to residences are limited to speeds of about a few hundreds kbps using xDSL modems because of the quality of the copper in the ground and cross-connections in the typical local loop. On the other hand, wireless broadband point-to-multipoint tele-communications platform facilitates the two-way transmission of voice, data, and video. The system typically operates in the 24 to 31 GHz bands, which, coupled with a high bandwidth capability in excess of 1.0 GHz, allows the communication of multimedia services with interactive facilities within a 3 mi (5 km) radius of a central data hub.

For the point-to-multipoint (PMP) architecture, the operator installs base stations around the market area that are very similar to traditional cellular systems,[5] and each cell contains a hub with multiple radio nodes equipped with sector antennas for PMP and directional antennas for PP connections. Base stations have antennas that transmit and receive on multiple sectors, and typical configuration is four sectors using 90° beamwidth antennas. The subscribers use antennas that are installed on their rooftops, pointing in the direction of the maximum signal strength from the base station. Similar to cellular system, frequencies are reused in neighbor base stations or sectors, as long as the reuse distances are defined so as to avoid interference. These systems provide the bandwidth on demand, which is achieved by use of ATM (and statistical multiplexing) as a transport mechanism. Since the bandwidth is shared with other users, bandwidth available per subscriber is reduced with every new subscriber using the system. For subscribers who demand fixed bandwidth availability, point-to-point system can be offered that will not share bandwidth with other subscribers. Figure 2.11 shows a typical system architecture based on the dynamic bandwidth allocation and ATM.

Millimeter-wave (MMW) characteristics dictate short-range line-of-sight propagation (rain attenuation) with minimal refraction and reduced interference. Line-of-sight between the hub site and all the customer sites is required, similar to classic point-to-point microwave systems. Therefore, standard microwave propagation and prediction methods are used for the system design. In addition, frequency coordination with other spectrum owners and PMP service providers has to be carefully planned.

In urban areas, only low LOS penetration rates are achievable, due to obstructions from buildings, vegetation, and other obstacles. One inexpensive option (at certain frequencies) to increase penetration might be the use of reflected and diffracted waves in the *non-line-of-sight (NLOS)* areas, if attenuation due to reflection or diffraction is smaller than a certain available system margin.[6] The use of reflected waves depends very much on the roughness of the surface under consideration. For very rough building walls with a standard deviation of the roughness close or more than $\lambda/2$, only reflected waves of grazing angles can be used. However, since roughness of building walls is often small compared to the millimeter wavelengths, reflection coeffi-

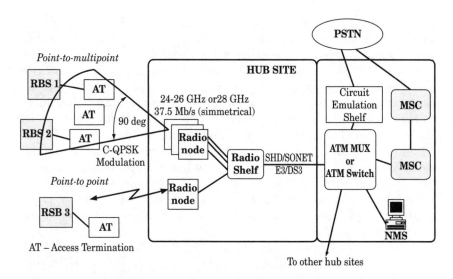

Figure 2.11 MW point-multipoint system architecture.

cients can be less than the system margin over a wide range of the angle of incidence.

PMP can use FDMA, TDMA, or spread-spectrum (DS or FH) over-the-air interface as well as frequency-division duplex (FDD) or time-division duplex (TDD). There is a debate about time-division duplex (TDD) vs. frequency-division duplex (FDD) in point-to-multipoint networks. While TDD requires a single channel for full-duplex communications, FDD system require a paired channel for communication, one for the downlink (hub to remote), and one for the uplink (remote to hub). In TDD, transmit/receive separation occurs in the time domain, as opposed to FDD, where it occurs in frequency domain. While FDD can handle traffic that has relatively constant bandwidth requirements in both directions, TDD better handles varying uplink/downlink traffic (bursty traffic—data, Internet) asymmetry by allocating time spent on up- and downlinks. TDD requires a guard time equal to the round-trip propagation delay between hub and remote units and increases with link distance. In FDD, sufficient isolation in frequency between the uplink and downlink is required, so FDD is simpler but less efficient solution.

2.5.2 Wireless Local Area Networks (WLANs)

Most corporate information systems and databases can be accessed remotely through the Internet protocol (IP) backbone, but the high bandwidth demand of typical office applications, such as large e-mail attachment downloading, often exceeds the transmission capacity of cellular/PCS networks. Mobile professionals are looking for a public access solution that can cover the demand for data-intensive applications and enable smooth online access to corporate data services. Wireless LAN access technology provides a perfect broadband complement for the operators' existing or new 3G services in an indoor environment. Seamless access to modern office tools is one of the most valuable assets for mobile business professionals today.

Public wireless LANs can handle large volumes of data at significantly lower costs, offer a migration path to higher speeds, and deliver additional capacity with pinpoint accuracy compared to leading 3G technologies. That is why 3G wireless network operators need public wireless LANs to serve the most demanding users in the most demanding locations.

Two standards dominate the WLAN marketplace; the IEEE 802.11b has been the industry standard for several years. Operating in the unlicensed portion of the 2.4-GHz radio frequency spectrum, it delivers a maximum data rate of 11 Mbps and boasts numerous strengths. Standard 802.11b enjoys broad user acceptance and vendor support. Many vendors manufacture compatible devices, and this compatibility is assured through the Wi-Fi certification program. Thousands of enterprise organizations that typically find its speed and performance acceptable for their current applications have deployed 802.11b technology. In the U.S.A., a number of wireless ISPs have emerged that are offering public access services using IEEE 802.11b equipment operating in the 2.4 GHz band. These providers are targeting public areas where business travelers may wish to access corporate intranets or the Internet, e.g., in hotels or coffee shops.[7] In some parts of Europe, there are now a number of service providers, both mobile operators and ISPs, offering wireless Internet services based on 802.11b technology in the 2.4-GHz band.

Another WLAN standard, IEEE 802.11a, operates in the uncluttered 5-GHz radio frequency spectrum. With a maximum data rate of 54 Mbps, this standard offers a fivefold performance increase over the 802.11b standard. Therefore, it provides greater bandwidth for particularly demanding applications. The IEEE ratified the 802.11a standard in 1999, but the first 802.11a-compliant products did not begin appearing on the market until December 2001. The 802.11a standard delivers a maximum data rate of 54 Mbps and eight non-overlapping frequency channels—resulting in increased network capacity, improved scalability, and the ability to create microcellular deployments without interference from adjacent cells.

Operating in the unlicensed portion of the 5 GHz radio band, 802.11a is also immune to interference from devices that operate in the 2.4 GHz band, such as microwave ovens, cordless phones, and Bluetooth (a short-range, low-speed, point-to-point, personal-area-network wireless standard).

The 802.11a standard is not compatible with existing 802.11b-compliant wireless devices. The 2.4-GHz and 5-GHz equipment can operate in the same physical environment without interference.

IEEE 802.11g is a high-performance standard and will deliver the same 54 Mbps maximum data rate as 802.11a, but it will operate in the same 2.4 GHz band as 802.11b.

Selecting between these technologies is not a one-for-one trade-off. They are complementary technologies and will coexist in future enterprise environments. Implementers must be able to make an educated choice between deploying 2.4-GHz-only networks, 5-GHz-only networks, or a combination of both. Organizations with existing 802.11b networks cannot simply deploy a new 802.11a network on 5-GHz access points (APs) and expect to have similar coverage with 802.11a 54-Mbps data rate as compared to 11 Mbps of data rate with 802.11b APs. The technical characteristics of both these bands simply do not allow for this kind of coverage interchangeability.

2.5.3 Bluetooth

Bluetooth technology allows for the replacement of the many proprietary cables that connect one device to another with one universal short-range radio link. For instance, Bluetooth radio technology built into both the cellular telephone and the laptop would replace the cumbersome cable used today to connect the two devices. Printers, PDAs, desktops, fax machines, keyboards, joysticks, and virtually any other digital device can be part of the Bluetooth system. However, beyond replacing the cables, Bluetooth radio technology provides a universal bridge to existing data networks, a peripheral interface, and a mechanism to form small private ad hoc groupings of connected devices away from fixed network infrastructures. Bluetooth is a *de facto* open standard for short-range digital radio.

Bluetooth is considered to be a point-to-multipoint system, although it can also be a point-to-point system, depending on the application. Bluetooth operates in the 2.4-GHz industrial, scientific, and medical (ISM) band, an unlicensed portion of the spectrum that is already well used and generally available in most parts of the world (Table 2.3). Not only do microwave ovens operate within this range, but so do other RF communications technologies, most notable of which are HomeRF and IEEE 802.11b.[8] Because this spectrum is unlicensed, even more uses for it are expected to develop in the future. As the band becomes more widely used, radio interference will increase. Bluetooth uses FHSS and is a shorter-range and lower-bandwidth technology than 802.11b, and it uses frequently changing, narrow bands over all channels. It is important to manage the concurrent operation of 802.11b WLANs and Bluetooth within the enterprise. Task Group 2 of the IEEE 802.15 Working Group is looking at the coexistence issues of IEEE 802.11b WLANs and Bluetooth.

TABLE 2.3   Bluetooth Frequency Bands

Area	Frequency band (GHz)	Number of channels
USA, Europe, and other countries	2.4000–2.4835	79
Spain	2.4450–2.475	23
France	2.4465–2.4835	23

To counter this interference, Bluetooth technology incorporates several techniques to provide robust linkages; for example, cyclical redundancy encoding, packet retransmission, and frequency hopping, which can occur up to 1,600 times per second.

2.6 References

1. ITU, *Handbook of Radiometeorology,* Geneva, 1996.
2. Mojoli, L. F. and Mengali, U., "Propagation In Line of Sight Radio Links" (Part II—Multipath Fading), *Telletra Review,* 1988.
3. Mojoli, L. F. and Mengali, U., "Propagation In Line of Sight Radio Links" (Part I—Visibility, Reflections, Blackout), *Telletra Review,* 1988.
4. Department of Commerce, "Federal Long-Range Spectrum Plan," prepared by Working Group 7 of the Spectrum Planning Subcommittee, September 2000.
5. Lehpamer, H., *Transmission Systems Design Handbook for Wireless Networks,* Norwood, MA: Artech House, 2002.
6. Hayn A., Jakoby R., "Radio Propagation Aspects for Digital Microwave Video Distribution System (MVDS) at 42 GHz," Institut fur Hochfrequenztechnik, 1998.
7. Mason Communications Ltd., "Spectrum Management Strategies for Licence-Exempt Spectrum: Final Report," London, England, November 2001.
8. Agilent Technologies, "Investigating Bluetooth™ Modules: The First Step in Enabling Your Device with a Wireless Link," Application Note 1333-2.

3

Microwave Link Design

3.1 Design Process Flowchart

Microwave link design is a methodical, systematic, and sometimes lengthy process that includes the following main activities:

- Loss/attenuation calculations
- Fading and fade margins calculations
- Frequency planning and interference calculations
- Quality and availability calculations

A preliminary fade margin is a result of the loss/attenuation calculations and is used for preliminary fade predictions in the fading calculation. If interference is present in the frequency-planning calculation, then the threshold degradation is included in the fade margin. The updated fade margin will become the effective fade margin used in the fading predictions. The results of the loss/attenuation and fading calculations will form the necessary input to the quality and availability calculations. The whole process is iterative and may go through many redesign phases before the required quality and availability are achieved (see Fig. 3.1).

3.2 The Loss/Attenuation Calculations

The loss/attenuation calculation is composed of three main contributions: propagation, branching, and "miscellaneous" (or other) losses. The propagation contribution comes from the losses due to the Earth's atmosphere and terrain—e.g., free-space as well as gas, precipitation

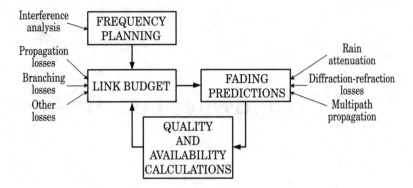

Figure 3.1 Microwave link design process.

(mainly rain), ground reflection, and obstacles. The branching contribution comes from the hardware required to deliver the transmitter/receiver output to the antenna—e.g, waveguides as well as splitters and attenuators. The "miscellaneous" contribution has a somewhat unpredictable and sporadic character, e.g., sandstorms as well as fog, clouds, smoke, and moving objects crossing the path. In addition, poor equipment installation and less than perfect antenna alignment (field margin) may give rise to unpredictable losses. The "miscellaneous" contribution normally is not calculated, but it can be considered in the planning process as an additional loss and then as part of the fade margin.

3.2.1 Propagation Losses

3.2.1.1 Free-space loss. Electromagnetic waves are attenuated while propagating between two geometrically separated points. The free-space path loss model is used to predict received signal strength when the transmitter and receiver have a clear, unobstructed line-of-sight path between them. The attenuation is inversely proportional to the square of distance and gives the free-space loss that stands for most of the total attenuation caused by wave propagation effects. The frequency and distance dependence of the loss between two isotropic antennas is expressed in absolute numbers by the following equation:

$$L_{FSL} = \left(\frac{4\pi d}{\lambda}\right)^2$$

where d = distance between transmit and receive antennas (km)
       λ = operating wavelength (m)

The Friis free-space path loss model expressed here is valid only for distances that are in the far field of the transmitter antenna. If D represents largest linear dimension of the antenna (diameter of the parabolic dish antenna), d represents the transmitter-receiver (T-R) separation distance, and λ represents the wavelength, then the following relationships define the far-field region:

$$d \gg D$$

$$d \gg \lambda$$

$$d \gg 2D^2/\lambda$$

Free-space loss is always present, and it is dependent on distance and frequency. The free-space loss between two isotropic antennas is derived from the relationship between the total output power from a transmitter and the received power at the receiver. After converting to units of frequency and expressing it in the logarithmic (decibel) form, the equation becomes

$$L_{FSL} = 92.45 + 20\log(f) + 20\log(d) \quad [\text{dB}]$$

where f = frequency (GHz)
d = line-of-sight (LOS) range between antennas (km)

Close inspection of the free-space path loss equation yields a relationship that is useful in dealing with link budget issues. Each 6-dB increase in EIRP equates to a doubling of range. Conversely, a 6-dB reduction in system losses (either by way of transmission line loss or on the transceiver end) translates into a doubling of range (the "6-dB rule"). This is not always so easy to accomplish, because the total path attenuation is also determined by other negative contributions, e.g., gaseous losses and rain.

3.2.1.2 Vegetation attenuation. LOS between stations is required for point-to-point microwave links. For an unexpected obstacle intercepting the Fresnel zone (e.g., growing vegetation), the additional loss can be calculated. High-resolution path profiles and careful site and path surveys are important to avoid unexpected obstacle attenuation. Vegetation is continuously growing, and the rate of growth is very important. It is important to include a provision for at least five years of vegetation growth.

Foliage losses at millimeter-wave frequencies are significant. An empirical relationship was developed (CCIR Report 236-2) that can predict the loss. For the case in which the foliage depth is less than 400 m, the loss is given by

$$L = 0.2\, f^{0.3} R^{0.6}\ \text{(dB)}$$

where f = frequency (MHz)
   R = depth of foliage (m)

This relationship is applicable for frequencies in the range 200 to 95,000 MHz. For example, the foliage loss at 40 GHz for a penetration of 10 m (which is about equivalent to a large tree or two in tandem) is about 19 dB. This is clearly a very serious attenuation and has to be considered or, if possible, avoided.

3.2.1.3 Gas absorption. Nitrogen and oxygen molecules account for approximately 99 percent of the total volume of the atmosphere. Since the absorption bands of nitrogen is located far from the radio-relay region of the spectrum, the atmosphere is considered to be composed of a mixture of two "gases": dry air (oxygen molecules) and water vapor (water molecules). The two absorption peaks present in the frequency range of commercial radio links are located around 23 GHz (water molecules) and 50 to 70 GHz (oxygen molecules). Specific attenuation (in dB/km) for water vapor and oxygen are separately calculated and then summed to give the total specific attenuation. The specific attenuation is strongly dependent on frequency, temperature, and the absolute or relative humidity (RH) of the atmosphere (see Fig. 3.2).

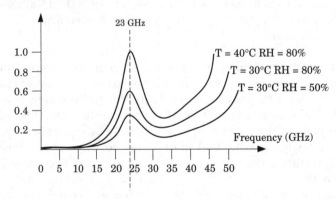

Figure 3.2 Gas attenuation versus frequency.

Many other atmospheric gases and pollutants have absorption lines in the millimeter bands (e.g., SO_2, NO_2, O_2, H_2O, and N_2O); however, the absorption loss is primarily due to water vapor and oxygen only.

3.2.1.4 Attenuation due to precipitation. Precipitation can take the form of rain, snow, hail, fog, and haze. In common for all of the above forms of precipitation is the fact that they all consist of water particles (haze can also consist of small solid particles). Their distinctions lie in the distribution of the size and form of their water drops.

Rain attenuation is, however, the main contributor in the frequency range used by commercial radio links. The main parameter used in the calculation of rain attenuation is rain intensity (rain rate), which is obtained from cumulative distributions. These distributions are the percentage of time for which a given rain intensity is attained or exceeded and are furnished for 15 different rain zones covering the entire Earth's surface.

The specific attenuation of rain is dependent on many parameters such as the form and size distribution of the raindrops, polarization, rain intensity, and frequency. The contribution due to rain attenuation is *not* included in the link budget and is used only in the calculation of rain fading. It is important to notice that rain attenuation increases exponentially with rain intensity (mm/h) and that horizontal polarization gives more rain attenuation than vertical polarization. Rain attenuation increases with frequency and becomes a major contributor in the frequency bands above 10 GHz.

3.2.1.5 Obstacle loss. Diffraction is the mechanism responsible for obstacle loss/attenuation. In fact, obstacle loss is also known in the literature as *diffraction loss* or *diffraction attenuation*. Depending on the shape, size, and properties of the obstacle, diffraction calculations can be cumbersome and time consuming. Since microwave paths normally require LOS, relatively simple methods for calculating the obstacle loss are currently employed. One powerful but simple method for calculation of obstacle loss is the single-peak method, which is based on the knife-edge approximation. This method can easily be extended to comprise the three most significant peaks inside the Fresnel zones. If a knife-edge approximation is considered, the values given in Fig. 3.3 are reasonable approximations. Having an obstacle-free 60 percent of first Fresnel zone gives 0 dB obstruction loss.

3.2.2 Ground Reflection

Reflection on the Earth's surface may give rise to multipath propagation. Depending on the path geometry, the direct ray at the receiver

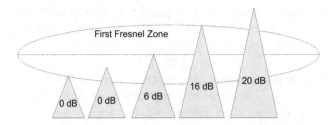

Figure 3.3 Obstacle losses and knife-edge approximation.

may be interfered with by the ground-reflected ray, and the *reflection loss* can be significant. Since the refraction properties of the atmosphere are constantly changing (*k*-value changes), the reflection loss varies (fading). The loss due to reflection on the ground is dependent on the total reflection coefficient of the ground and the phase shift. Figure 3.4 illustrates the signal strength as a function of the total reflection coefficient. The highest value (A_{Max}) of signal strength is obtained for a phase angle of 0°, and the lowest value (A_{Min}) is for a phase angle of 180°.

The reflection coefficient is dependent on the frequency, grazing angle (angle between the ray beam and the horizontal plane), polarization, and other ground properties. The grazing angle of radio-relay paths is very small—usually less than 1°. It is strongly recommended to avoid ground reflection, which can be achieved by "shielding" the path against the indirect ray. For large grazing angles, the difference between vertical and horizontal polarization is substantial.

Changing the antenna heights can move the location of the reflection point. This approach, usually known as the *hi-lo technique,* forces the reflection point to move closer to the lowest antenna by affecting the height of the higher antenna. The grazing angle increases, and the

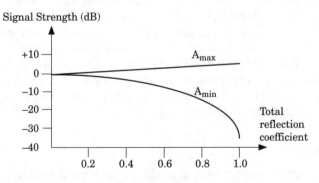

Figure 3.4 Signal strength versus reflection coefficient.

path becomes less sensitive to k-value variations. Space diversity also provides good protection against reflection, and it is usually applied for paths over open water surfaces.

Obviously, on many paths, particularly at higher frequencies, it is difficult to obtain an accurate estimate of the effective surface reflection coefficient because of various uncertainties such as the surface conductivity, surface roughness, and so on, and the degree of subjectivity currently needed to obtain a calculation. The calculation procedure may only be a rough guide in such situations to help identify problem paths or to help choose one path from another, even if this possibility exists in the first place.

The contribution resulting from *reflection loss* is *not* automatically included in the link budget. However, when reflection cannot be avoided, the fade margin may be adjusted by including this contribution as "additional loss" in the link budget.

3.3 Fading and Fade Margins

Fading is defined as the variation of the strength of a received radio carrier signal due to atmospheric changes and/or ground and water reflections in the propagation path. Four fading types are normally considered when planning for radio-relay paths.

- Multipath fading, which is divided into

  - *Flat fading*

  - *Frequency-selective fading*

- *Rain fading*

- *Refraction-diffraction fading* (k-type fading)

All fading types are strongly dependent on the path length and are estimated as the probability of exceeding a given (calculated) fade margin.

3.3.1 Multipath Fading

Various clear-air fading mechanisms caused by extremely refractive layers in the atmosphere must be taken into account in the planning of links of more than a few miles in length; *beam spreading* (commonly referred to as *defocusing*), *antenna decoupling, surface multipath,* and *atmospheric multipath.* Most of these mechanisms can occur by themselves or in combination with each other.

Multipath fading (flat or frequency selective) is the dominant fading mechanism for frequencies lower than approximately 10 GHz. A reflected wave causes a phenomenon known as *multipath*, meaning that the radio signal can travel multiple paths to reach the receiver. Typically, multipath occurs when a reflected wave reaches the receiver at the same as the direct wave that travels in a straight line from the transmitter. Two cases are possible.

- If the two signals reach the receiver *in phase,* then the signal is amplified. This is known as an *upfade.* Upfades can also occur when the radio wave is trapped within an atmospheric duct. As can be seen from the following formula, higher upfades are possible for longer paths:

$$Upfade_{max} = 10 \log d - 0.03d \text{ (dB)}$$

Path length d is in kilometers and, for the 50-km path, maximum upfade can be up to 16.6 dB.

- If the two waves reach the receiver *out of phase,* they weaken the overall received signal. If the two waves are 180° apart when they reach the receiver, they can completely cancel each other out so that a radio does not receive a signal at all. A location where a signal is canceled out by multipath is called a *null* or *downfade.*

Smooth surfaces, such as a body of water, a flat stretch of earth, or a metal roof, reflect radio signals. In Fig. 3.5, the body of water reflects a wave that cancels out the direct signal and brings down the radio link.

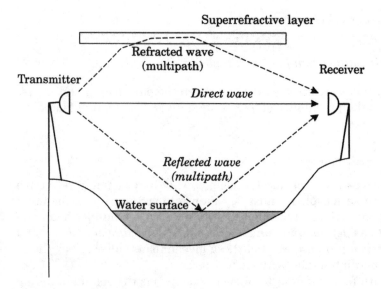

Figure 3.5 Multipath fading.

Multipath fading was also observed on Earth-to-space links for elevation angles below about 5°; however, the phenomenon is more commonly encountered on terrestrial links.[1] All ray components on an Earth-space link usually traverse similar vertical refractive conditions. While there may be some signal-level fading or enhancement due to beam spreading or convergence, the similar impact on all components tends to make the multipath phenomenon less prevalent than on terrestrial links where rays traveling along different heights may encounter distinctly different refractive conditions along their entire lengths.

Some important facts about multipath fading are as follows:

- Multipath fading is normally more active over bodies of water (lakes, sea, and so forth) than over land. It is common practice on over-water paths to use a low-high antenna pair to move any multipath reflections out of the antenna main beam.

- It is important to avoid ground reflection. Multipath fading is more likely on paths across flat ground than on paths over rough terrain. Horizontal paths give most flat fading.

- Multipath fading is normally most active during early and later summer (late spring and early autumn). Calm weather favors atmospheric stratification, and that gives multipath fading. The fading season is defined as the so-called "worst fading month," usually a summer month.

- In radio links of typical length and in temperate climates, multipath activity lasts approximately three months, so the yearly fading season length is one-fourth of that measured in the worst month. When operating in tropical climates with very long hop lengths, multipath activity may last up to six months with intensity comparable to that of the worst month.

- A rule of thumb is that multipath fading, for radio links having bandwidths less than 40 MHz and path lengths less than approximately 30 km, is described as being flat instead of frequency selective.

- Increasing path inclination reduces the effects of flat fading. Reducing path clearance will reduce the effect of flat fading, because the risk of multipath propagation is decreased; however, this technique may increase the risk for refraction-diffraction fading.

- On over-water paths at frequencies above about 3 GHz, it is advantageous to choose vertical polarization over horizontal polarization. At grazing angles greater than about 0.7°, a reduction in the surface reflection of 2 to 17 dB can be expected over that at horizontal polarization.

Links should be sited to take advantage of rough terrain in ways that will increase the path inclination (sometimes referred to as the *high-low technique*). This approach should be conducted jointly with more specific efforts to use shielding from terrain to reduce the levels of surface reflection. Where towers are already in place, antenna height at one end of the path could be reduced to accomplish this as long as the clearance rules are satisfied.

Experimental evidence indicates that, in clear-air conditions, fading events exceeding 20 dB on adjacent hops in a multi-hop link are almost completely uncorrelated. This suggests that, for microwave systems with large fade margins, the outage time for a series of hops in tandem is approximately given by the sum of the outage times for the individual hops.

As digital radio continues to operate at higher data rates and with more complex modulation, the need for multipath fade testing during installation and routine maintenance increases. Measurement printout is required for comparison to specified performance documentation during radio line up or for comparison between different radios.

3.3.1.1 Flat fading. A flat fading is just another way of describing a fade (or reduction in input signal level) where all frequencies in the channel of interest are equally affected. Flat fading implies barely noticeable variation of the amplitude of the signal across the channel bandwidth. Flat fading is dependent on path length, frequency, and path inclination. In addition, it is strongly dependent on the geoclimatic factor (temperature/pressure variations), which is the factor that accounts for the refraction properties in the atmosphere, antenna altitudes, and the type of terrain.

If necessary, the flat fade margin of a link can be improved, including using larger antennas, a higher-power microwave transmitter, lower-loss feed line, and splitting a longer path into two shorter hops in several ways.

3.3.1.2 Frequency-selective fading. Frequency-selective fading implies amplitude and group delay distortions across the channel bandwidth produced by the multipath nature of the transmission media.[2] It particularly affects medium- and high-capacity radio links (>32 Mbps). The sensitivity of digital radio equipment to frequency-selective fading can be described by the signature curve of the equipment. The signature parameter definitions and specification of how to obtain the signature are given in Recommendation ITU-R F.1093.

The basic principle behind the construction of signature curves of equipment is to split the radio path between a transmitter and a receiver into two signals, one direct and the other reflected/refracted.

There is a time delay between both signals before they are finally combined in the receiver. In the laboratory, the split is simulated by a time-delayed signal (the reflected/refracted signal). The equipment signature is a measure of the receiver's capability to suppress the time-delayed signal. The signature is therefore the level of the signal that is necessary to obtain a certain BER (currently referred to as 10^{-3} and/or 10^{-6}) in the presence of an interfering signal with a pre-defined delay and it is measured in the laboratory.

Measurements are normally performed at the IF stage (typically 70 MHz). The phase difference between the direct and indirect signals causes a notch (dip) at one of the frequency positions inside the spectrum bandwidth. By changing the phase difference between the direct and indirect signals, the *notch frequency* will change inside the bandwidth. At every notch frequency, the signal is attenuated until the threshold for a specific BER (10^{-3} and/or 10^{-6}) is exceeded. The final diagram (Fig. 3.6) will illustrate the sensitivity (currently known as the *notch depth* and expressed in dB) of the receiver as a function of the notch frequency. The difference between the highest and lowest notch frequency is the *signature bandwidth*, expressed in megahertz.

It is easier to understand and measure the performance of a microwave radio under multipath receptions when considering just one direct signal and one indirect signal having a different amplitude and relative time delay. Since the amplitude of the reflected signal can be lower or higher than the direct signal, *minimum or nonminimum phase group can be obtained*. When the amplitude of the direct signal is higher than that of the indirect signal, the notch is called a *minimum-phase notch*. Conversely, when the delayed signal has higher amplitude than the direct signal, the notch is a *nonminimum phase notch*. Because a receiver can respond differently to these types of

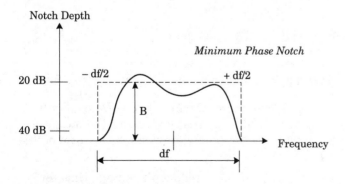

Figure 3.6 Microwave radio signature curve.

notches, it is important to test the radio under both minimum and nonminimum phase conditions. In general, nonminimum is more severe than minimum phase dispersive fading, but, under most conditions, the direct signal typically is stronger (minimum phase notch).

The M-curve or outage signature is a plot of the minimum phase notch depth vs. the notch frequency at which the BER of the link starts to exceed a certain threshold. When the notch depth at a given frequency is greater than the value on the curve, the BER is unacceptable. The W-curve is a plot of nonminimum phase notch vs. notch depth. In either case, the curve can be used to calculate a dispersive fade margin that is equal to the area enclosed by the curve and the horizontal frequency axis.

The *dispersive fade margin (DFM),* a value expressed in decibels, is the measure of a receiver's ability to resist dispersive fading. Digital microwave radio manufacturers measure and provide the dispersive fade margin from the typical (measured) fading signatures for their digital radio receivers. To measure the DFM of a digital radio receiver, manufacturers simulate multipath fading conditions either in the field or at the factory. W. D. Rummler of Bell Laboratories has developed a simplified three-path model of multipath propagation and has shown that 6.3 ns is approximately the delay time measured on real microwave links in the U.S.A. The dispersive fading lab simulation will work at either the carrier (RF) frequency or the intermediate frequency (IF). Most manufacturers of multipath simulation systems incorporate IF in their designs because of relative ease of design and better accuracy at lower frequencies.

The dispersive fade margin is defined as the average depth of multipath fade that causes an outage, independent of thermal and interference fade margins.

$$DFM = 17.6 - 10\log\left[\frac{2(\Delta f)e^{-B/3.8}}{158.4}\right]$$

where Δf = signature width of the equipment
B = notch depth of the equipment

It is important to remember the following facts about selective fading:

- If Δf and B are not available, the user can define DFM explicitly as well. In Europe, signature width is usually given for the microwave radio; in North America, it is commonly provided in the form of dispersive fade margin.

- Increasing the output power so as to reduce the outage time for selective fading does not give any improvement. It only increases the flat fading or reduces the thermal noise power received without having any influence on the effects (amplitude and group delay distortions across the channel) of selective fading.

- Modern digital microwave radios are very robust and immune to dispersive (spectrum-distorting) fade activity. Only a major error in path engineering (wrong antenna size or misalignment) over the high-clearance path could cause dispersive fading problems.

3.3.2 Rain Fading

The principal gaseous absorption is by oxygen and water vapor. Oxygen loss is negligible for frequencies up to about 50 GHz and will be neglected in this analysis. The first and best known effect of rain is that it attenuates the signal. The attenuation is caused by the scattering and absorption of electromagnetic waves by drops of liquid water. The scattering diffuses the signal, while absorption involves the resonance of the waves with individual molecules of water. Water vapor absorption is highly dependent on the frequency as well as the density of the water vapor (absolute humidity, gm/m^3). Water vapor absorption can be significant for long paths (>10 km). Loss has a local maximum at 22 GHz and a local minimum at about 31 GHz. Absorption increases the molecular energy, corresponding to a slight increase in temperature, and this results in an equivalent loss of signal energy.

Attenuation is negligible for snow or ice crystals, in which the molecules are tightly bound and do not interact with the waves. The extent of the attenuation due to rain is primarily a function of the form and the size distribution of the raindrops. Rain fading starts increasing noticeably at about 10 GHz and, for frequencies above 15 GHz, rain fading is normally the dominant fading mechanism.

Rain events are statistically predictable with reasonable accuracy if short-integration or instantaneous rain measurements are available. Models that are based on measured cumulative distributions of rain events are currently employed in the prediction of the probability that a certain fade margin will be exceeded. The model estimates the time (normally expressed in percentage of a year) during which a given fade depth (fade margin) is exceeded. Next, the result is converted to worst-month statistic. The concept of *worst month* for a certain specific value of the worst month is defined as that month with the highest probability of exceeding that specific value.

Other forms of precipitation (snow, hail, fog, and haze) do not affect radio-relay links as much as rain events and are considered negligible. For example, at 23 GHz, a 5-mi-long microwave link will have additional attenuation due to a very dense fog of only about 0.7 dB. Snow covering antennas and radomes, the so-called ice coating, can result in different types of problems such as increased attenuation and deformation of the antenna's radiation diagram.

The rain rate enters into this equation, because it is a measure of the average size of the raindrops. When the rain rate increases (i.e., it rains harder), the raindrops are larger, and thus there is more attenuation. Rain models differ principally in the way the effective path length L is calculated. Two authoritative rain models that are widely used are the Crane model and the ITU-R (CCIR) model.

Heavy rainfall, usually in cells accompanying thunderstorm activity and weather fronts, has a great impact on path availability above 10 GHz. Rain outage increases dramatically with frequency and then with path length. Fading due to rain attenuation is described empirically from link tests and point rainfall data. Location variation is based on selected point rainfall data and radar reflectivity data accumulated around the world. Ten- to 15-minute duration fades to over 50 dB have been recorded on an 18-GHz, 5-km (3-mi) path, for example, and increased outage at 23 GHz can require a 2-to-1 reduction in path length compared to 18 GHz for a given availability.

Much is known about the qualitative aspects, but the problems faced by the microwave transmission engineer remain formidable. To estimate probability distribution, instantaneous rainfall data is needed. Unfortunately, the available rainfall data is usually in the form of a statistical description of the amount of rain that falls at a given measurement point over various time periods. The total annual rainfall in an area has little relation to the rain attenuation for the area. In some cases, the greatest annual rainfalls are produced by long periods of steady rain of relatively low intensity at any given time. Other areas of the country, with lower annual rates, experience thunderstorms and frontal squalls, which produce short-duration rain rates of extreme intensity. The incidence of rainstorms of this type determines the rain rates for an area, and thus the high-frequency microwave link's long-term path outage time and "unavailability." Even the rain statistics for a day or an hour have little relationship to rain attenuation. A day with only a fraction of an inch/centimeter of total rainfall may have a path outage due to a short period of concentrated, extremely high-intensity rain. Another day with several inches/centimeters of total rainfall may experience little or no path attenuation, because the rain is spread over a long time period or large area. The predicted annual outage may not occur for years and then accumulate over a single rainy season for a long-term average.

The worst rain outages occur during the heaviest thunderstorms. The gulf coast area from Florida to New Orleans has the most severe thunderstorms in the U.S. As a result, rain outages in microwave systems are most severe in the southeastern U.S. Microwave path lengths must be reduced in these areas to maintain the path availability. Cities such as Seattle, Washington, and Vancouver, Canada, also receive a large amount of rain. However, there are few thunderstorms, so rain outage is less severe, and longer path lengths are feasible.

In the design of any engineering system, it is impossible to guarantee the performance under every conceivable condition. One sets reasonable limits based on the conditions that are expected to occur at a given level of probability. For example, a bridge is designed to withstand loads and stresses that are expected to occur in normal operation and to withstand the forces of wind and ground movement that are most likely to be encountered. However, even the best bridge design cannot compensate for a very powerful tornado or an earthquake of unusual strength. Similarly, in the design of a microwave communications link, one includes a margin to compensate for the effects of rain at a given level of availability.

3.3.3 Refraction-Diffraction Fading

In the real world, the k-factor varies with time and location in accordance with complex physical interactions involving the refractivity gradient (dn/dh) in the lowest part of the atmosphere and other mechanisms as detailed in the propagation "P series" of ITU Recommendations. An important objective in planning terrestrial microwave link systems is to ensure that outages resulting from these variations are extremely rare events; thus, system fade margins, linked to error performance and availability objectives, of the appropriate order are implemented to ensure that this is so. Accordingly, to take account of the statistical nature of radio wave propagation, the application of appropriate propagation prediction models is necessary.

Refraction-diffraction fading, also known as k-type fading, is characterized by seasonal and daily variations in the Earth-radius factor k. For low k values, the Earth's surface becomes more curved, and terrain irregularities, man-made structures, and other objects may intercept the Fresnel zones. For high k values, the Earth's surface gets close to a plane surface, and better LOS (lower antenna heights) is obtained. The probability of refraction-diffraction fading is therefore indirectly connected to obstruction attenuation for a given value of Earth-radius factor. Since the Earth-radius factor is not constant, the probability of refraction-diffraction fading is calculated based on cumulative distributions of the Earth-radius factor.

3.3.4 Interference Fade Margins

To accurately predict the performance of a digital radio path, the effect of interference must be considered. Interference in microwave systems is caused by the presence of an undesired signal in a receiver. When this undesired signal exceeds certain limiting values, the quality of the desired received signal is affected. To maintain reliable service, the ratio of the desired received signal to the (undesired) interfering signal should always be larger than the threshold value.

Interference into a digital radio will degrade the receiver threshold and result in a lower effective fade margin, thus producing excessive bit errors or frame losses as the radio fades near threshold. In normal, nonfaded conditions, the digital signal can tolerate high levels of interference; however, to protect short-term performance and hop reliability, it is critical to control interference in deep fades.

Adjacent-channel interference fade margin (AIFM) (in decibels) accounts for receiver threshold degradation due to interference from adjacent channel transmitters in one's own system. AIFM is applicable only to frequency diversity and multiline (N+1) protection systems and is not used as long as the minimum allowable frequency separations for a particular radio are maintained. These minimum frequency separations are included in the manufacturer's radio specifications.

Interference fade margin (IFM) is the depth of fade to the point at which RF interference degrades the BER to 1×10^{-3}. It is affected by the frequency congestion, directivity of the interfering and victim system antennas, and so on.

The actual IFM value used in a path calculation depends on the method of frequency coordination being used.[3] There are two widely used methods: the "C/I" and "T/I" methods. The C/I method is an older one, developed originally to analyze interference cases into analog radios. Frequencies are selected such that the calculated carrier-to-interference (C/I) ratio for all interfering transmitters is less than some objective value. The objectives are listed in National Spectrum Manager's Association (NSMA) interference objective tables.

In the new T/I method, threshold-to-interference (T/I) curves are used to define a curve of maximum interfering power levels for various frequency separations between interfering transmitter and victim receivers as follows:

$$I = T - (T/I)$$

where I =maximum interfering power level (dBm)
T =radio threshold for a 10^{-6} BER (dBm)
T/I =threshold-to-interference value (dB) from the T/I curve for the particular radio

For each interfering transmitter, the receive power level in dBm is compared to the maximum power level to determine whether the interference is acceptable. This is done instead of calculating C/I values and comparing them to the published objectives. The T/I curves are based on the actual lab measurements of the radio.

3.3.5 Composite Fade Margin

Composite fade margin (CFM) is the fade margin applied to multipath fade outage equations for a digital radio link. The complete expression for describing the CFM for a digital microwave radio is given by

$$CFM = TFM + DFM + IFM + AIFM$$

$$= -10\log(10^{-TFM/10} + 10^{-DFM/10} + 10^{-IFM/10} + 10^{-AIFM/10})$$

where TFM = flat fade margin, the difference between the normal (unfaded) RSL and the BER = 1×10^{-3} DS1 loss-of-frame point.

DFM = dispersive fade margin, provided by the radio manufacturer from measurements using the W-curve. It is affected by the complexity of the digital modulation scheme and the types and effectiveness of the adaptive amplitude and/or baseband time domain equalization (if any) used.

IFM = interference fade margin.

$AIFM$ = adjacent-channel interference.

These four fade margins are power added to derive the CFM. The longer the link, the more critical the above factors become, since the system gain and composite fade margin determine the range and the reliability performance of a radio under various fading conditions. Often, only the dominant terms (namely, the flat and the dispersive fade margin) are included.

3.4 Microwave Link Multipath Outage Models

A major concern for microwave system users is how often and for how long a system might be out of service. Various statistical models and analysis methods have been developed to predict and measure the outage and availability over a period of time. Performance prediction (related to propagation effects) principally depends on the assessment of two main propagation mechanisms: multipath fading and/or attenuation due to rain. Multipath fading typically gives rise to short outages and thus has the most impact on error performance. For modeling multipath fading, propagation prediction methods have been derived that estimate the probability of single-frequency fading. Rain attenuation events typically give rise to outage durations greater than 10 sec, therefore directly influencing availability of systems operating in the bands above about 10 GHz. The prediction of rain outages is possible through the application of rainfall intensity statistics to modeling methods for rain attenuation.

An outage in a digital microwave link occurs with a loss of DS1 frame sync (OOF) for more than 10 sec. DS1 frame loss typically occurs when the BER increases beyond 1×10^{-3}.

$$Outage\ (Unavailability)\ (\%) = (SES/t) \times 100$$

where t = time period (expressed in seconds)
SES = severely errored second, a state defined as any 1-sec period containing a BER of 1×10^{-3} or greater, often accompanied by an out-of-frame DS signal, but with no service disruption

Availability is expressed as a percentage as follows:

$$A = 100 - Outage\ (Unavailability)$$

A digital link is unavailable for service or performance prediction/verification after a ten consecutive BER > 1×10^{-3} SES outage period.

3.4.1 Vigants North American Multipath Outage Model

This Vigants model is also widely used in ITU-R regions and can be found in computer programs as CCIR Rep.338 using KQ geoclimatic factors. The average probability of multipath outage in Vigants North American model (1975) over a long period of time (for example, one year) due to fading is given by

$$P = 2.5 \times 10^{-6}\ c f D^3\ 10^{-CFM/10}$$

where P = one-way probability of outage due to all multipath fade activity during the fade season (%)
f = frequency (GHz)
D = path length (miles)
CFM = composite fade margin
c = climate/terrain factor (4 over water and humid climate, 1 for average terrain and climate, 0.25 for mountains and dry climate)

or

$$c = a \times (50/w)^{1.3}$$

where $a = 2$ for humid climate, 1 for average climate, 0.5 for dry climate

w = average terrain roughness (over a range of 20 to 140 ft, 50 ft being "normal") extracted from the path profile

The tilting terrain of a high/low path computes to $w > 140$ ft for $c = 0.52$ maximum in humid areas using this method.

Path length, D, is raised to the "power of 3" and hence has a significant effect on the multipath outage and, in turn, on the performance objectives in terms of availability expectations. Multipath outage is highly dependant on the terrain roughness factor, and mountainous paths exhibit least multipath.

Outage time in seconds per year (one-way) can be calculated as follows:

$$Outage = T_0 \times P \text{ (sec/year)}$$

and fade duration, T_0 is expressed as a fraction of a year is given by

$$T_0 = 8 \times 10^6 \ t/50 \text{ for yearly availability}$$

where t = average annual temperature in degrees Fahrenheit ($35°F \leq t \leq 75°F$)

Multipath fading is a warm-weather phenomenon. This calculation assumes an average three months (8×10^6 sec) fade season, although in reality the fade season is proportional to average annual temperature in the area and can vary between 2.1 and 4.6 months.

3.4.2 ITU-R Multipath Outage Model

A method for predicting the single-frequency (or narrowband) fading distribution suitable for large fade depths in the average worst month in any part of the world[4] and for detailed link design is given as follows:

$$P_0 = Kd^{3.2}(1 + |\varepsilon_p|)^{-0.97} \times 10^{0.032f - 0.00085 h_L} \text{ (\%)}$$

where f = frequency (GHz)

h_L = altitude of the lower antenna (i.e., the smaller of h_e and h_r)

and the geoclimatic factor K is obtained from the following equation (if measured data for K are not available):

$$K = 10^{-3.9 - 0.003dN_1s_a - 0.42}$$

where dN_1 = point refractivity gradient in the lowest 65 m of the atmosphere not exceeded for 1 percent of an average year
s_a = area terrain roughness

The term dN_1 is provided on a 1.5° grid in latitude and longitude in Recommendation ITU-R P.453. The correct value for the latitude and longitude at path center should be obtained from the values for the four closest grid points by bilinear interpolation. The data are available in a tabular format and are available from the Radiocommunications Bureau (BR).

Term s_a is defined as the standard deviation of terrain heights (m) within a 110 × 110 km area with a 30 s resolution (e.g., the Globe "gtopo30" data). The area should be aligned with the longitude such that the two equal halves of the area are on each side of the longitude that goes through the path center. Terrain data are available from the Internet, and web address is provided by the BR.

If a quick calculation of K is required for planning applications, a fairly accurate estimate can be obtained from

$$K = 10^{-4.2 - 0.0029dN_1}$$

From the antenna heights h_e and h_r (meters above sea level), calculate the magnitude of the path inclination $|\varepsilon_p|$ (mrad) using the following expression:

$$|\varepsilon_p| = \frac{|h_r - h_e|}{d} \text{ (mrad)}$$

where d = the path length (km)

The method shown above is used for small percentages of time, does not make use of the path profile, and can be used for initial planning, licensing, or design purposes.

A second method, used for all percentages of time, is suitable for all fade depths and employs the method for large fade depths and an interpolation procedure for small fade depths. The method used for predicting the percentage of time that any fade depth is exceeded combines the deep fading distribution and an empirical interpolation procedure for shallow fading down to 0 dB. A detailed description of a method for all percentages of time is beyond the scope of this book but can be found in recommendation ITU-R P.530-10.

3.5 Quality and Availability Calculations

The main purpose of the quality and availability calculations is to set up reasonable quality and availability objectives for the microwave path. The entire procedure can be structured in five general steps (see Fig. 3.7).

An appropriate network model is selected in step 1, along with the selection of quality and availability objectives for the corresponding portions and sections of the network model given in step 2. In step 3, the quality and availability parameters are calculated and compared to the objectives in step 4. If the objectives are not met (step 5), appropriate network parameters (antenna size, antenna height, output power, channel arrangements, polarization, and so on) are changed, and the quality and availability parameters are recalculated as illustrated in step 3. The procedure is continued in step 4 as an iterative process.

For predicting the quality of a microwave radio link, the following fading mechanisms are usually considered:

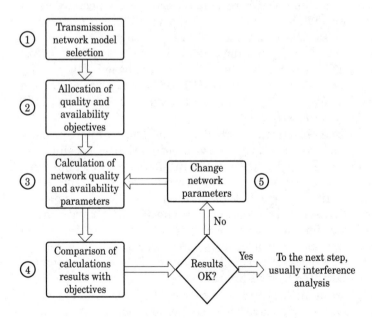

Figure 3.7 Quality and availability objectives.

- Flat fading due to multipath propagation

- Selective fading due to multipath propagation

- Fading due to rain

- Fading due to refraction-diffraction in the atmosphere, also known as k-type fading

Multipath fading (flat and selective) is assumed to cause fast-fading events. The major contribution is ES, ESR, and BBE. Rain and refraction-diffraction fading are assumed to cause relatively slow-fading events, giving UAT, ES, and BBE. The major contribution is, however, unavailable time.

3.5.1 G.821 and G.826

In digital transmission technology, any bit received in error (a *bit error*) may deteriorate transmission quality. It is obvious that quality will decrease with an increasing number of erroneous bits. Therefore, the ratio of the number of errored bits referred to the total number of bits transmitted in a given time interval is a quantity that can be used to describe digital transmission performance. The quantity called *bit error ratio (BER)* is a well known error performance parameter.[5]

Bit error ratio can be measured only if the bit structure of the evaluated sequence is known. For this reason, bit error ratio measurements are mostly performed out of service, using a well defined *pseudorandom bit sequence (PRBS)*. In practice, the PRBS replaces the information sent in service. Quality parameters defined by Rec. ITU-T G.821 and applied by ITU-R recommendations are BER based. The quality parameters are *errored second ratio (ESR)* and *severely errored second ratio (SESR)*. The availability parameter is *available time ratio (ATR)* or *unavailable time ratio (UATR)*.

One of the prime objectives of G.826 was to define all performance parameters in such a way that in-service estimation is possible. In-service detection of errors in digital transmission is possible, however, using special error detection mechanisms (error detection code, EDC) which are inherent to certain transmission systems. Examples of those inherent EDCs are cyclic redundancy check (CRC), parity check, and observation of bit interleaved parity (BIP). EDCs are capable of detecting whether one or more errors have occurred in a given sequence of bits—the block. It normally is not possible to determine the exact number of errored bits within the block.

The basic philosophy of G.826 is based on the measurement of errored blocks, thus making in-service error estimation possible. Block errors are processed in a similar way as bit errors, i.e., the term *block error ratio* is defined as the ratio of the number of errored blocks referred to the total number of blocks transmitted in a given time interval. The quality parameters are *errored second ratio (ESR), severely errored second ratio (SESR),* and *background block error ratio (BBER).* The availability parameter is *available time ratio (ATR)* or *unavailable time ratio (UATR).*

Some useful error performance definitions are as follows:

- *Errored block (EB)*—a block in which one or more bits are in error.

- *Errored second (ES)*—a one-second period with one or more errored blocks or at least one defect.

- *Errored second ratio (ESR)*—the ratio of ES to total seconds in available time during a fixed measurement interval.

- *Severely errored second (SES)*—a one-second period that contains over 30 percent errored blocks or at least one defect. SES is a subset of ES.

- *Severely errored second ratio (SESR)*—the ratio of SES to total seconds in available time during a fixed measurement interval.

- *Background block error (BBE)*—an errored block not occurring as part of an SES.

- *Background block error ratio (BBER)*—the ratio of background block errors (BBE) to total blocks in available time during a fixed measurement interval. The count of total blocks excludes all blocks during SESs.

Consecutive severely errored seconds (CSES) may be precursors to periods of unavailability, especially when there are no restoration/protection procedures in use. Periods of CSES persisting for T seconds (2–10) (some network operators refer to these events as *failures*) can have a severe impact on service, such as the disconnection of switched services. Error performance should be evaluated only while the time the path is in the available state.

Measurement of BER and BLER yields comparable results for small BERs, and, for some specific error models, it is possible to calculate BER from a BLER. It is the drawback of this procedure that error models describe the situation found in practice only imperfectly and may be strongly media dependent. Therefore, the result of such a calculation is not very reliable.

A *hypothetical reference path (HRP)* as defined by Rec. ITU-T G.826 is the whole means of digital transmission of a digital signal of a specified rate, including the path overhead (where it exists) between equipment at which the signal originates and terminates. An end-to-end HRP spans a distance of 27,500 km. The portion of interest is usually the national portion of the HRP subdivided in three classes: access, short-haul and long-haul (see Fig. 3.8).

For the purposes of G.826, the boundary between the national and international portions is defined to be at an *international gateway (IG)*, which usually corresponds to a cross-connect, a higher-order multiplexer, or a switch (N-ISDN or B-ISDN). IGs are always terrestrially based equipment, physically resident in the terminating (or intermediate) country.

A large number of devices (test equipment, transmission systems, collecting devices, operating systems, software applications) are currently designed to estimate the G.821 or M.2100[6] parameters ESR and SESR at bit rates up to the fourth level of the PDH. For such devices, the G.826 parameters ESR and SESR may be approximated using the G.821 criteria, but an approximation of BBER is not possible, since the block-based concept and the BBER parameter are not defined in the Recommendation G.821.

3.5.2 Quality and Unavailability Objectives

The quality and unavailability objectives for the all portions in the hypothetical reference path (HRP) have to be accomplished concurrently. These objectives shall account for effects caused by fading, interference, and other sources of performance degradation.

Block allocation is used in the access and short-haul portions, while in the long-haul portion it is a combination of block allocation and length-related allocation. Rain fading and diffraction-refraction fading (k-type fading) give unavailability whereas multipath (flat and frequency selective) fading gives ESR, SESR, and BBER. Unavailability is the dominating dimensioning factor for frequencies above 15 GHz, whereas quality is the dominating and dimensioning factor for frequencies below 10 GHz (8 GHz in some countries). There is, however, a frequency range between 10 and 15 GHz where both quality and availability might be comparable, and all the mechanisms have to be considered.

Distance-based allocation in the access and short-haul is not supported by any ITU-R recommendation based on Rec. ITU-T G.826,

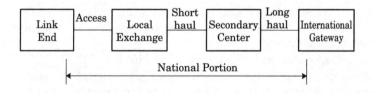

Figure 3.8 Hypothetical reference path—national portion.

G.827, and G.828 (portions of the network model are not length defined). Distance-based allocation distributes the objectives equally along a radio link (chain of paths). This is not always effective, because "difficult paths" normally require more objective apportionment than "easy paths." Unavailability due to hardware failure is not path-length related, but unavailability due to radio wave propagation (rain and refraction-diffraction fading) can be strongly length dependent.

Transmission network requirements for the reliability (yearly outage) in the microwave networks in North America run from 30 sec (99.9999 percent) for *high-reliability links* to 26 min (99.995 percent) for *single MW hops* carrying less-important traffic. Sometimes, these numbers are different and could be proposed by the customer himself; it is important to keep in mind that, by definition, *higher objective numbers lead to a more expensive MW network.*

The ITU-T recommendations G.801, G.821, and G.826 define error performance and availability objectives. The objectives for digital links are divided into separate grades: high, medium, and local grade. The medium grade has four quality classifications. The following grades are usually used in wireless networks:

- Medium grade Class 3 for the access network
- High grade for the backbone network

3.6 Rain Attenuation Calculations

The most-used method today for calculating rain attenuation was developed by the International Telecommunications Union Radio Section. However, there are other methods, such as the Crane method (used mainly in the United States) and methods more suitable for specific radio climates, such as wet tropical and equatorial areas. There are differences between the two most popular models, the ITU terrestrial and Crane models, which produced slightly different estimates of the long-term fade probability. Variability is inherent in the estimation of path fading because empirical equations have been derived from measurement points around the world over limited periods of time. Rain attenuation, which is the dominant fading mechanism for millimeter wave paths, is based on nature, which can vary from location to location and from year to year.

It is obvious that the uncertainty of either model or, alternatively, the short-term expectation of fade, is quite large. The uncertainty, as measured by the estimated attenuation standard deviation, is greater than 30 percent and tends to overshadow arguments about the accuracy of the methods. The uncertainty is a result of variations from year to year and location to location. Location to location within a rain zone, we also find a high estimated attenuation standard deviation. In addition, worst-month predictions forecast much higher fade depths than the basic annual predictions. An underestimate of the required margin to compensate for a given probability of rain outage results in a system that does not meet link availability requirements.

If availability guarantees are accepted in system contracts, cost incentives may have to be paid for underdesigning a system. Overestimations of the required margin result in the overdesign of systems, which results in unnecessary system costs.

The Crane models (after Robert K. Crane) are popular for space-Earth links but also have terrestrial models. There are three versions of the Crane models. The *global Crane model* was developed in 1980. In 1982, the *two-component Crane model* was developed, which used a path-integrated technique. A volume cell contribution and a debris contribution for a path were computed separately and added to provide a link calculation. As a refinement of the two-component model, the *revised two-component model* was introduced in 1989, which includes spatial correlation and statistical variations of rain within a cell. All these models are described in Crane's book.[7]

A rain event consists of small "volume cells" of intense rain rate within much larger "debris regions" with a lower rain rate. The dimensions of these areas are inversely related to rain rate. Volume cells are quite small, generally less than 5.0 km^2, with the average volume cell area of about 3.0 km^2 over the range of rain rates 20 to 40 mm/hr, which is often of interest as design targets for microwave systems. This could be represented as a rectangle of about 1.6 × 3 km diameter (rain cells are usually not circular).

Calculation by the ITU model is straightforward by scaling the 0.01 percent rain rate and by using an effective path-length reduction factor to account for the cellular nature of heavy rainfall. Mean cumulative distributions of rainfall zones are defined geographically in ITU-R 837.[8] The Crane rainfall zones are defined differently from the ITU zones, with more defined zones in the U.S. than the ITU zones.

Currently, the ITU-R provides Rec. ITU-R P.530-xx[*] for calculating the average annual, one minute averaged, rain fade distribution experienced by a terrestrial link. The prediction method refers to Rec. ITU-R P.837-2 for calculating the average annual, one minute averaged, rain rate distribution and to Rec. ITU-R P.838 for the specific attenuation-rain rate relationship. Rec. ITU-R P.530-xx also provides some guidance for extending these models for the performance of a simple link to predict the performance of more complex links such as multihop links and links utilizing diversity.

---

[*]Presently used and the latest revision of this model is ITU-R-P.530-10-.

The ITU has recommended a calculation method for terrestrial systems, ITU-R P.530-xx, and for space-to Earth-links, ITU-R P.618. These models take into account a distance reduction factor to account for the cellular nature of storms and have improved since the original CCIR version.

In most cases, the Crane models predict higher rain attenuation or, in other words, they are more conservative than the ITU model.

More detailed information on attenuation due to hydrometeors other than rain is given in Recommendation ITU-R P.840.

3.6.1 Crane model

The method is used to predict the attenuation by rain on terrestrial propagation path. As a first step, the rain climate region of the endpoints of the path has to be determined. Calculating the outage is an iterative process that is based on the calculation of the attenuation.

For $x \le d \le 22.5$ km,

$$A = \alpha R_p^\beta \left[\left(\frac{e^{\mu\beta d} - 1}{\mu\beta} \right) - \left(\frac{b^\beta e^{c\beta x}}{c\beta} \right) + \left(\frac{b^\beta e^{c\beta d}}{c\beta} \right) \right] \quad \text{(dB)}$$

For $d < x$,

$$A = \alpha R_p^\beta \left(\frac{e^{\mu\beta d} - 1}{\mu\beta} \right) \quad \text{(dB)}$$

Values of μ, x, b, and c can be calculated as follows:

$$\mu = \ln(b e^{cx})/x$$

$$b = 2.3\, R_p^{-0.17}$$

$$c = 0.026 - 0.03 \ln(R_p)$$

$$x = 3.8 - 0.6 \ln(R_p)$$

where d = path length in kilometers

R_p = rain rate in millimeters per hour, determined from Crane tables

e = natural logarithm ($\log_{2.71828}$)

α, β = regression coefficients obtained from the table and interpolated if necessary

First, it is necessary to determine the rainfall rate required to produce an attenuation equal to the thermal fade margin. After that, determine the percent of the year this rain rate is exceeded. This value represents the annual two-way rain outage time for the path and is calculated using the rainfall statistics for the required geographic region.

If $d > 22.5$ km, the rain attenuation is calculated using $d_0 = 22.5$ km and a modified probability of occurrence p_1.

$$p_1 = p(22.5/d)$$

3.6.2   ITU-R Model

The prediction procedure outlined here is considered valid in all parts of the world, at least for frequencies up to 40 GHz and path lengths up to 60 km. The following simple technique may be used for estimating the long-term statistics of rain attenuation:

1. Obtain the rain rate $R_{0.01}$ exceeded for 0.01 percent of the time (with an integration time of 1 min). If this information is not available from local sources of long-term measurements, an estimate can be obtained from the information given in Recommendation ITU-R P.837.

2. Compute the specific attenuation, Y_R (dB/km) for the frequency, polarization, and rain rate of interest using Recommendation ITU-R P.838.

3. Compute the effective path length, d_{eff}, of the link by multiplying the actual path length d (in kilometers) by a distance factor r as follows:

$$d_{eff} = dr$$

$$r = \frac{1}{1 + \dfrac{d}{d_0}}$$

where, for $R_{0.01} \leq 100$ mm/h,

$$d_0 = 35e^{-0.015R_{0.01}}$$

For $R_{0.01} > 100$ mm/h, use the value 100 mm/h in place of $R_{0.01}$.

4. An estimate of the path attenuation exceeded for 0.01 percent of the time is given by

$$A_{0.01} = Y_R d_{eff} = Y_R dr \quad \text{(dB)}$$

5. For radio links located in latitudes ≥30° (north or south), the attenuation exceeded for other percentages of time p in the range 0.001 to 1 percent may be deduced from the following power law:

$$\frac{A_p}{A_{0.01}} = 0.12p^{-(0.546 + 0.043\log_{10}p)}$$

This formula has been determined to give the following factors: 0.12 for $p = 1.00$ percent, 0.39 for $p = 0.1$ percent, 1.00 for $p = 0.01$ percent, and 2.14 for $p = 0.001$ percent.

6. For radio links located at latitudes below 30° (north or south), the attenuation exceeded for other percentages of time p in the range 0.001 to 1 percent may be deduced from the following power law:

$$\frac{A_p}{A_{0.01}} = 0.07p^{-(0.855 + 0.139\log_{10}p)}$$

This formula has been determined to give the following factors: 0.07 for $p = 1.00$ percent, 0.36 for $p = 0.1$ percent, 1.00 for $p = 0.01$ percent and 1.44 for $p = 0.001$ percent.

The attenuation A in the above equations is set to the fade margin and the equation is solved for p.

ITU-530 clearly states that the equations are valid only in the range from 1 to 0.001 percent. On many practical links, this range will be exceeded, especially on short links with high fade margins.

7. If worst-month statistics are required, calculate the annual time percentages p corresponding to the worst-month time percentages p_w using climate information specified in Recommendation ITU-R P.841. The values of A exceeded for percentages of the time p on an annual basis will be exceeded for the corresponding percentages of time p_w on a worst-month basis.

Both ITU-530-xx and Crane methods calculate the annual probability of rain outage. The annual rain outage probability is translated to worst-month rain outage probability as follows:

$$p_w = 2.85p_a^{0.87}$$

3.6.3 Comparison of ITU and Crane Models

The comparison in Table 3.1 of predicted attenuation is provided for places where the ITU and Crane zones overlap. ITU zone M does not correspond to a Crane zone very well and is not included in the com-

TABLE 3.1   Rain Attenuation Comparison at 99.99 Percent Availability for a 3-km Path

ITU zone/Crane zone	Units	E/F	D/C	K/D2	N/E
Rain rate ITU/Crane	(mm/hr)	22/22	19/29	42/47	95/91
ITU-R 530	(dB)	10.8	14.3	22.3	39.2
Crane global	(dB)	13.2	17.2	25.7	45.9
Crane 2-component	(dB)	13.6	18.4	28.8	52.0
Crane revised T-C	(dB)	12.4	20.0	26.9	51.3

parison. Crane D2 and E are irregular through M, with Crane E extending from Florida to Northern Alabama and up to South Carolina. Listed in the table are attenuation values for the same locations using the ITU zone for ITU calculations and the corresponding Crane zone for Crane calculations.

All of the Crane models predict a larger attenuation than the ITU model; however, this difference is also about the same as the difference between various Crane models.

It is interesting that ITU zone D is northern California, with less rainfall, and E is southern California, with more rainfall. The corresponding Crane zones are C for northern California and F for southern California, and they predict just the opposite intensity, which is actually correct.

For prediction of rain fade attenuation using the ITU 530-xx standard, rain rate at the 0.01 percent exceeded level for the zone of interest is required, plus frequency, path length, and attenuation factors from ITU-R 838. Other percentages are calculated using only the 0.01 percent value. The ITU model consists of simple equations, whereas all of the Crane models, on the other hand, require solution of a number of complex equations to obtain the path-averaged rain rate and a representation of the path profile by exponential functions.

Crane's formulas are used to make predictions about link availability for radio and radar installations. One shortcoming is encountered when trying to extend his result to other than surface to surface and surface to satellite links. Crane's "derivations" are overly empirical, sometimes hard to understand, and difficult to extend.

Users can apply in their calculations either ITU or Crane models, whichever makes them more comfortable. Their choice of model may depend more on their institutions' prejudices and traditions or their customers' "comfort level" with either model than on the requirements for accuracy. It is also important to keep in mind that the more stringent (conservative) method does not necessarily have to be more accurate. On the other hand, more conservative method of calculation can lead to more expensive network design.

3.6.4 Reducing the Effects of Rain

The most common reason for preferring a lower frequency is the susceptibility of bands above 10 GHz to rainfall attenuation. Although fades caused by rain cells are occasionally observed at lower frequencies (10 to 20 dB fades at 6 GHz have been recorded, even in North America), this type of fade generally causes outages only on paths above 10 GHz. The outages are usually caused by blockage of the path by the passage of rain cells (e.g., thunderstorms) perhaps 4 to 8 km (2.5 to 5 mi) in diameter and 5 to 15 min in duration on the path. Such fading exhibits slow, erratic level changes, with rapid path failure as the rain cell intercepts the path. Things to keep in mind in connection with rain attenuation fades are as follows:

- Multipath fading is at its minimum during periods of heavy rainfall with well aligned dishes, so the entire path fade margin is available to combat the rain attenuation (wet-radome loss effects are minimized with shrouded antennas).

- When permitted, seldom-used *crossband diversity* is very effective. In this case, the lower frequency path is stable (affected only by multipath fading) during periods when the upper frequency path is obstructed by rain cells.

- Neither space diversity nor in-band frequency diversity provides improvement against rain-attenuation fade outage. During a rainstorm, both antennas in a space diversity system, and all frequencies in a frequency diversity system, fade together.

- Intense rain showers that cause extreme attenuation occur in cells of small dimensions, so the probability that two or more links have deep fade events at the same time is very small, and it strongly depends on the geometry of the network. Route diversity with paths separated by more than about 8 km (5 mi) can be used successfully. To develop efficient route diversity methods or site diversity strategies for cellular mobile systems, deep knowledge of the spatial and time correlation statistic is essential.

- Increased fade margin is of some help in rainfall attenuation fading; margins as high as 45 to 60 dB, some with *automatic transmitter power control (ATPC)*, have been used in some highly vulnerable links for increased availability.

- Increasing the fade margins, shortening path lengths, using the lower frequency band, and increasing antenna sizes are the most readily available tools for reducing the per-hop annual rain outage in a given area.

- Because raindrops are oblate rather than spherical, attenuation tends to be greater for horizontally polarized signals than for vertically polarized signals. Vertical polarization is far less susceptible to rainfall attenuation (40 to 60 percent) than are horizontal polarized frequencies.

When long-term attenuation statistics exist at one polarization [either vertical (V) or horizontal (H)] on a given link and are expressed in decibels, the attenuation for the other polarization over the same link may be estimated through the following simple relation:

$$335A_V + A_VA_H - 300A_H = 0$$

Among the indirect consequences of precipitation, deep fades were experienced in winter months during periods of no precipitation at all. It turned out that these were consequences of reflections due to ice-layers on top of the snow. These can reflect millimeter waves, causing similar multipath effects as over the water propagation.

3.7 Improving the Microwave System

3.7.1 Hardware Redundancy

3.7.1.1 Hot standby protection. Equipment failure normally results in long interruptions (unavailability), and hardware redundancy often can be the only option of improving the total unavailability figure of a radio-link system. The simplest form of hardware redundancy is that of a standby system operating in parallel with the active system (1+1), and it takes over whenever the on-line (active) system fails. *Monitored hot standby* is preferred, and the term means that the functions of the standby component should work properly and optimally whenever required. A switching device connects the standby and the active system, and the term *monitored* implies electronic control/supervision.

Each RF channel requires two frequencies (transmit and receive). All transmit frequencies are in one half of the band, and all receive frequencies are in the other half. Frequencies are normally assigned so that all frequencies transmitting from a site are either in the high half or the low half of the band.

Hardware redundancy improves only the unavailability rate due to hardware failure; it does not affect the unavailability due to propagation effects, and it requires more equipment. Consequently, it is more expensive than a nonredundancy system.

The most important parameters in the calculation of hardware failure (unavailability) are the *mean time between failure (MTBF)* and the *mean time to repair (MTTR)*.

MTTR includes the time necessary to detect, report, locate, and repair the failure. Considering that the MTTR of nonredundant systems can be somewhat large in some applications (e.g., microwave systems in remote areas), redundancy is recommended if high availability (low unavailability) is required and short MTTR cannot be accomplished.

Unavailability is more suitable to radio-link planning than availability. The main reason for employing unavailability is that unavailability contributions can be added together when calculating the total unavailability of a system. The availability of a system should not be mixed with reliability. *Availability* means the probability of finding a system in operation when it is needed, whereas the *reliability* is the probability of a system operating as intended during a certain time interval and under certain conditions.

3.7.1.2 Multichannel and multiline protection. If the microwave system has more than one channel, protection can be achieved in two different ways. Multichannel protection system will have hot standby protection for every channel. In a multiline protection system, one standby channel will serve as a protection for N working channels. Multiline protection (N+1) requires an additional RF channel to use as a protection channel and, under normal working conditions, does not carry any traffic.

In other words, hot standby transmitters and receivers independently protect all working channels in a multichannel system at the same time. Multiline protection can only protect one working channel at a time. As a result, the equipment reliability is significantly better for a multichannel system but is also more expensive.

3.7.2 Diversity Improvement

Under normal conditions, only one propagation path exists between the transmitting and receiving antennas of a well designed LOS microwave link. *Multipath fading* occurs whenever a low-level reflected signal is radiated out of phase and, at a reduced level, is combined with the desired signal at the victim receiver (ground reflections). Usually, large variations in temperature and humidity (atmospheric refractions) accompany multipath fading, which is characterized by deep, fast, frequency-selective signal attenuation over a certain time period.

One method of limiting signal degradation caused by multipath propagation is to apply one or more of the three different diversity techniques (frequency, space, or angle diversity) or a combination of these. Since, at a given moment in time, each frequency in the bandwidth is affected differently, dispersive fading usually results, and diversity systems can be effective in stabilizing these conditions to a minimum when used with continuous, in-band power combiners in digital microwave applications. The most common forms of diversity in LOS links are frequency and space diversity (Fig. 3.9), although angle diversity is also used in rare occasions. Diversity does not help to overcome rain-fading events, only multipath and refraction-diffraction fading.

Methods are available for predicting outage probability and diversity improvement for space, frequency, and angle diversity systems, and for systems employing a combination of space and frequency diversity. The probability of outage for a diversity system is

$$U_d = \frac{U_{nd}}{l_d}$$

where U_d = one-way probability of outage for a diversity path
$\quad\quad I_d$ = diversity improvement factor

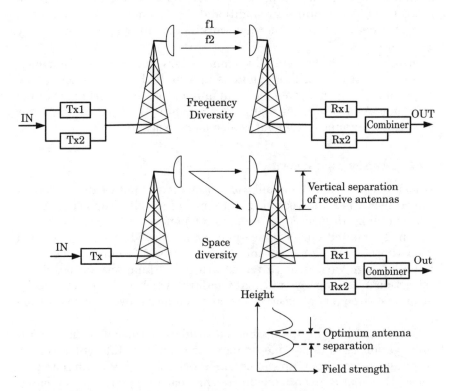

Figure 3.9 Frequency and space diversity.

The degree of improvement afforded by all of these techniques depends on the extent to which the signals in the diversity branches of the system are uncorrelated. For narrowband analog systems, it is sufficient to determine the improvement in the statistics of fade depth at a single frequency. For wideband digital systems, the diversity improvement also depends on the statistics of in-band distortion. The diversity improvement factor differs for each type of diversity—space, frequency, hybrid, and so forth.

Above 3 GHz, I_{sd} is nearly always higher (better) than I_{fd} and is therefore selected unless the diversity spacing, D_f, exceeds about 5 percent (300 MHz in the 6-GHz band, for example). Below 3 GHz, I_{fd} is usually larger than I_{sd} and is therefore selected, assuming that spectrum-governing bodies will allow the use of frequency diversity.

3.7.2.1 Space diversity. Space diversity is the most commonly used diversity option against multipath fading. It is typically used on long paths, shorter paths in poor propagation areas, and over-water paths to protect against surface reflections. Space diversity is the simultaneous transmission of the same signal over a radio channel by using two or more antennas for reception and/or transmission. When using space diversity, the improvement obtained depends on the extent to which the signals in the two diversity branches of the system are uncorrelated. The antennas are therefore physically separated on a tower or mast, and the distance between the antennas (antenna separation) at the receiver or transmitter is such that the individual signals are assumed uncorrelated (see Fig. 3.10). Hitless or errorless switches are used for switching the signal between the main and diversity branches.

The appropriate spacing of antennas in space diversity systems is determined by three factors:

1. To keep clearance of the lower antenna as low as possible (within the clearance so as to minimize the occurrence of surface multipath fading)

2. To achieve specified space diversity improvement factor for overland paths

3. To minimize the chance that the signal on one diversity antenna will be faded by surface multipath when the signal on the other antenna is faded

The space-diversity improvement factor (I_{sd}) for the flat fading component of the fade margin is proportional to the square of the antenna separation. At the same time, I_{sd} for the dispersive component of the

Figure 3.10 Space diversity microwave system over the water, Porto Vallarta, Mexico.

fade margin is independent of the vertical antenna separation greater than 10 ft (3.3 m). In extreme situations (e.g., very long over-water paths), it may be necessary to employ three-antenna diversity configurations. Angle diversity can be combined with space diversity to further enhance performance if desired, in which case the space-diversity antennas are tilted to provide the additional angle-diversity enhancement.

3.7.2.2 Frequency diversity. In a *frequency diversity* system, the radio at each terminal contains redundant transmitter-receiver pairs and simultaneously transmits the same signal over two or more radio frequency channels located at the same frequency band. Each transmitter operates on a different RF channel, and both transmitters are energized and, similarly, each receiver operates on a different RF channel but identical to the corresponding transmitter at the far end. When equipment failure or path fading affects an RF channel to the point at which the signal is degraded, a decision is made at the receiver end of the link as to which RF channel is placed on line. Another form of frequency diversity is *crossband diversity,* in which the RF bearers are in different frequency bands.

One great drawback to frequency diversity is the inefficient use of the available frequency spectrum, and it is prohibited in most countries. In the U.S.A., the FCC limits frequency diversity to encourage spectrum conservation [FCC Part 101.103(c)]. The rules state that frequency diversity transmission will not be authorized in these services if the operator cannot prove that the required communications cannot practically be achieved by other means. In the U.S.A., where frequency diversity is deemed to be justified on a protection channel basis, it will be limited to one protection channel for the bands 3700 to 4200, 5925to 6425, and 6525 to 6875 MHz, and a ratio of one protection channel for three working channels for the bands 10,550 to 10,680 and 10,700 to 11,700 MHz. In the bands 3700 to 4200, 5925 to 6425, and 6525 to 6875 MHz, no frequency diversity protection channel will be authorized unless there is a minimum of three working channels, except that, if there is a proof that a total of three working channels will be required within three years, a protection channel may be authorized simultaneously with the first working channel. A protection channel authorized under such exception will be subject to termination if applications for the third working channel are not filed within three years of the grant date of the applications for the first working channel.

The federal government bands are administered separately by the National Telecommunications and Information Administration (NTIA) and are not subject to FCC restrictions on frequency diversity or minimum loading. NTIA regulations allow 1+1 frequency diversity systems in the federal government bands.

3.7.2.3 Hybrid diversity. *Hybrid diversity (HD)* is an enhancement (SD+FD) of space diversity that uses frequency diversity (when permitted). Hybrid diversity is the most effective of all of the diversity arrangements and is preferred in difficult propagation areas and in space-limited mountaintop, urban area, and other sites that are restricted to single antennas. It is possible to compute frequency and hybrid diversity improvements for links in regions where regulatory rules or waivers so permit. The hybrid diversity improvement factor, I_{hd}, is derived either from the space I_{sd} or frequency diversity improvement factor I_{fd} described above.

The higher T/R frequencies must always be assigned to the upper antenna at the space diversity (usually the lower elevation) end of each hybrid diversity link for optimum performance.

Using microwave radio systems below 10 GHz, hop lengths of about 50 km (30 mi) usually are achieved without problems, utilizing standard diversity techniques. When longer distances have to be spanned, however, the problems of propagation increase dramatically. In the early 1990s, Siemens installed one of the longest over-the-water microwave hops for the Mexican telephone company, Telmex. This hop was installed across the Gulf of California in the so-called Baja California region, and was 160 km (100 mi) long. Many "classic" models for estimating the fading behavior of a radio hop assume one reflection point and therefore two superimposed waves at the receiver location. This basic assumption does not apply under extreme conditions such as this one.

To meet quality requirements under these very problematic propagation conditions, a customized solution and a number of unusual technical measures were implemented.

- Four large 4.6-m parabolic antennas at each side were used.

- Selection of the best of eight received signals at any time was made (four antennas with two different frequencies).

- Custom-designed, very low-loss channel branching filters, using waveguide technology without circulators, were implemented.

- Increased frequency spacing, achieving greater efficiency of frequency diversity reception, was used.

Theory and practice show that a higher number of receiving antennas increase the probability of receiving a strong and less distorted signal.

3.7.2.4 Angle diversity. *Angle diversity (AD)* has been used in line-of-sight digital microwave links since the mid 1980s and in troposcatter links since the 1950s. This method used to be popular when digital radios were less robust, but today AD antennas are assigned mostly where installation constrictions (roof mounts problems, space, aesthetics, tower loading, and so on) prohibit SD and thus justify these less-effective, more-costly dishes.

The angle diversity antenna is a single dish with two feeds vertically offset by about 1° (the smaller, the better). Angle diversity is most effective when path outages are dominated by dispersive fade activity (i.e., the dispersive fade outage approaches or exceeds flat fade outage). Depending on path geometry and climatic conditions, angle diversity improvements of perhaps 20 or even much more are achieved.

Optimum angle-diversity improvements are achieved through an antenna alignment procedure that matches the antenna size and alignment to the path and its climatic characteristics. Angle diversity dishes require a more exacting, long-term alignment procedure than that for space diversity and nondiversity antennas.

3.7.2.5 Route diversity. When a point in a network requires high availability, it is often more spectrum efficient to install multiple links to combat rain than to use higher transmit power on a single connection. The simplest case of diversity is a point connected to a network by two independent links that may operate at different frequencies and polarizations. In the case of *route diversity*, the links connect to two different points in the network and so operate along different paths. This scenario is described by the length, frequency, and polarization of each link and the azimuth angle between the two links where they converge at the point being serviced and the integration time of fade measurements. The availability is increased because, when one link has failed due to rain attenuation, the other may still be operational. The increase in availability is quantified using the *diversity improvement factor*. The higher the correlation of rain attenuation on the two links, the lower the gain from using diversity, as the failure of one link becomes an increasingly better predictor of the failure of the other link. This correlation is determined by the geometry of the link system and the spatial-temporal statistics of rain intensity.

3.7.2.6 Media diversity. Important microwave links can be protected by using completely different media. Most commonly used are fiber-optic systems. In addition, the opposite is true, and sometimes fiber-optic systems are protected by high-capacity microwave links. These solutions are very expensive and should be considered only in exceptional cases.

3.7.3 Antireflective Systems

Antireflection refers to an antenna arrangement technique for reducing the effects of multipath fading. If two identical antennas at either (or both) the receiving site and transmitting site are used, they can be arranged so that their resulting radiation pattern has a null in the direction of the reflection point.[9,10] In addition, this antenna arrangement gives an extra 3-dB gain to the direct ray.

If the direct ray arrives horizontally, the two antennas are set in the vertical plane with a spacing of h meters (see Fig. 3.11). The output of each antenna is connected into a hybrid through two feeders, which are cut to length so that the direct received signals add in phase at the hybrid output. To trim the phase for optimum performance, a phase shifter is used in one of the feeders. Because of the different path lengths traversed by the reflected signals in reaching each of the two antennas, their phases will not add in phase. The combination of both reflected signals may, under certain conditions, lead to their cancellation.

The practical limitations with antireflective systems are

- Changes in k factor will change the angle between the direct and reflected ray.

- There may be movement of the antenna due to wind.

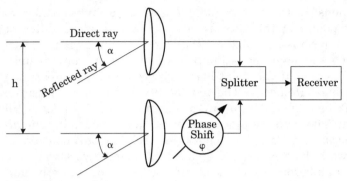

Cancellation condition

$$\varphi = \frac{2\pi}{\lambda} h \sin\alpha = (2n + 1)\pi$$

Figure 3.11 Antireflective system.

- Temperature changes affect the length of the feeders.
- They exhibit narrow tolerances in carrier frequency.
- A hybrid and delay network, sensitive to temperature variations, should be installed in the equipment room.
- Antireflective system should not be used for low values of α.

If α is fixed by the geometry of the link, the distance h between antennas is as follows:

$$h = \left(\frac{\lambda}{2}\right)(2n + 1)\left(\frac{1}{\sin\alpha}\right)$$

For example, at a frequency of 6.15 GHz ($\lambda = 0.048$ m) and $\alpha = 1°$, $h = 1.4$ m (for $n = 0$), and $h = 4.19$ m (for $n = 1$). The distance between antennas will depend on their diameter.

3.7.4 Repeaters

In cases here a direct microwave path cannot be established (i.e., no line of sight) between two points, it is possible to establish a path by using a repeater. The function of such a repeater is to redirect the beam so as to pass the microwave beam around or over the obstacle (e.g., a building or hill). The main requirement is that there be a clear line of sight between the repeater and both sides of the microwave link. This could be an active repeater (two microwave radios connected back to back) if distances are long, or a passive repeater if distances are relatively short.

3.7.4.1 Active repeaters. An *active MW repeater* site contains two complete microwave radio terminals (connected back to back), antennas, waveguides or coax cables, and other components, and it is a much more costly solution than the passive repeater described in the next chapter. It requires an enclosure for the equipment, power plant, an antenna-mounting structure of some kind, and so on. The best way to avoid use of active microwave repeater sites is to carefully plan and execute the microwave network design and strategically place sites in such a way that they all have a LOS with at least one other site. Active repeaters are used not only in the case of obstructed LOS but also when terminal stations are to far for one microwave hop. Two sites 100 mi apart can be typically connected using three, four, five, or even more microwave repeater stations, depending on the frequency, terrain (LOS), type of equipment (radios and antennas) used, and so forth.

3.7.4.2 Passive repeaters. There are two types of passive repeaters in use. One consists of two parabolic antennas connected back to back through a short piece of transmission line. The other, more commonly used, is a plane reflector, flat "billboard"-type metal reflector that acts as a microwave mirror. The effectiveness of a billboard-type passive repeater is an inverse function of the product of the lengths of the two paths rather than the sum of their lengths, as one might suppose. Thus, it is highly desirable to keep one of the paths very short (a few hundred meters, if possible).

Passive repeaters are used to change the direction of the radio-relay signal so as to overcome obstacles in an otherwise direct line of sight between two microwave (radio-relay) stations.[11] They are also employed when the cost as compared with an active repeater is too high. The use of passive repeater is not only less expensive than an active repeater, the operation cost is also substantially reduced, as neither power nor access roads are required.

Plane reflectors reflect the microwave signal in the same way as mirrors reflect light. A *single-plane reflector (single billboard)* consists of a flat reflexive surface that changes the ray direction to avoid the obstacle (see Fig. 3.12). The performance of this setup is given by the reflector surface (height and width) and the angle α, formed between the incident and reflected ray. For single billboards, things are simple if the billboard is in the far field of both antennas. In that case, the antenna gains and the billboard gains are independent and do not interact with each other. Then, we simply calculate a total path loss (which

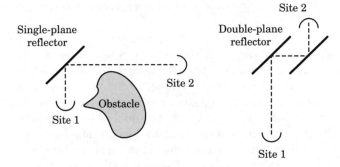

Figure 3.12 Single- and double-plane reflectors.

is the sum of the two antenna gains and the two-way gain of the reflector) to arrive at the end-to-end path loss through the reflector.

$$\text{Net path loss} = A1 + A2 - G1 - G2 - G3$$

where $A1$ = short-leg attenuation (dB)
$A2$ = long-leg attenuation (dB)
$G1, G2$ = antenna gains (dB)
$G3$ = free-space, two-way gain of a single passive billboard (dB)

Passive repeater gain calculations can be complicated when the billboard is in the near field of one of the parabolic antennas. A rule of thumb formula to determine the near field boundary is

$$D = 2 \times f \times B^2$$

where D = near-field zone distance (ft)
f = frequency (GHz)
B = antenna diameter (ft)

In that case, net path loss is

$$\text{Net path loss} = A1 + A2 - G1 - G2 - (G3 - K)$$

The correction factor K (dB) can be calculated using empirical formulas and graphs, and its value will depend on the parabolic dish diameter, frequency, and the distance between the parabolic antenna and the passive repeater. Its value is usually between 0.2 and 1.6 dB.

Double-plane reflectors are variations of the one-plane reflector solution. They operate by changing the direction of the radio-relay signal twice. This application is limited due to costs and difficulty of aligning the radio link.

Back-to-back antennas work like ordinary active repeaters, but without generating radio frequency signals. Back-to-back antennas are practical when the reflection angle is large, so they currently are employed in suburban environments, whereas plane reflectors, due to their large size, are used in rural applications. Back-to-back antennas consist of two antennas connected by a waveguide or a short coaxial cable (see Fig. 3.13). The total path length, the two additional antenna gains, and the loss introduced by the waveguide are included in the link budget. Therefore, back-to-back antennas introduce considerable losses, and they are less efficient than plane reflectors.

More information on passive repeater engineering, as well as other antenna mounting structures, and corresponding standards, can be found in Refs. 11 and 12.

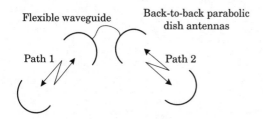

Figure 3.13 Back-to-back antennas.

3.8 References

1. Segal B.,"Multipath propagation mechanisms deduced from tower-based meteorological measurements," Comm. Research Centre, Ottawa, Canada, 2000.
2. Mojoli, L. F. and Mengali, U., Propagation in Line of Sight Radio Links (Part II – Multipath Fading), *Telletra Review,* 1988.
3. TIA, Telecommunications Systems Bulletin, TSB-10-F, Interference Criteria for Microwave Systems, 1994.
4. Recommendation ITU-R P.530-10, Propagation Data and Prediction Methods Required for the Design of Terrestrial Line-of-Sight Systems, 2001.
5. ITU-T Recommendation G.826, Error Performance Parameters and Objectives for International, Constant Bit Rate Digital Paths at or above the Primary Rate, 02/99.
6. ITU-T Recommendation M.2100, Performance Limits for bringing into Service and Maintenance of International PDH Paths, Sections and Transmission Systems, 07/95.
7. Crane, R. K., *Electromagnetic Wave Propagation through Rain,* New York: John Wiley & Sons, 1996
8. ITU-R PN.837, Characteristics of Precipitation for Propagation Modeling.
9. Mojoli, L. F. and Mengali, U., Propagation in Line of Sight Radio Links (Part I – Visibility, Reflections, Blackout), *Telletra Review,* 1988.
10. Alcatel Telspace, "Note on Digital Radio Link Calculation (Medium and High Capacity)," SP/LN1/A/04.96.
11. GTE Lencurt Inc., Engineering Considerations for Microwave Communications Systems, 1970.
12. Valmont Microflect, catalog 41.

Planning the Microwave Network

So far, we have talked about microwave-link design challenges. Even if we were able to design a perfect microwave link, it would never work in isolation from other links in the microwave system we are designing or far enough away to be immune to the influence of other (external) radio communication systems. Microwave network planning and microwave network design focus on the challenges of making the network work properly in an environment that is full of signals from other users of the RF spectrum.

The main difference between microwave network planning and microwave network design is that one cannot build a network based on the outcome and deliverables of the network planning process, whereas network design results are used during the deployment phase to actually build the network. Microwave network planning is a set of preliminary activities and information gathering used only to determine a need and a feasibility (a *feasibility study* can also be a separate phase) of the microwave network build-out and to consider other options such as building a fiber-optic system or leasing lines from the existing wireline operators. Microwave network planning analyzes the traffic model to arrive at network capacity and dimensions and a microwave network architecture, and also to make a preliminary equipment choice. In addition, it will determine budget, time lines, and the work force required for the successful completion of the project and help to get approvals to actually finalize the detailed design and build the network.

4.1 The Microwave Network Planning Process

To plan a network effectively, a network planner needs to have a complete understanding of the whole network, understand the business

objectives of the operator, and be able to respond to business requirements effectively. At all stages, the emphasis should be on designing a
simple network architecture. This will be beneficial in terms of deployment and network management and will provide flexibility to allow
for easy network expansion. During the first stage of the transmission
network planning process, a few initial questions need to answered
with regard to areas such as economics, the area topology, the existing
network, and the services the customer wishes to offer. The following
are a few examples:

- Who is the operator, what economic resources does he have, and
  what kind of services is the operator going to offer?
- Are we planning many years ahead or just dealing with today's de-
  mand?
- Is there an existing network, and what is a spare capacity available?
- Are we expanding an existing network or designing a new one?
- What are the requirements for reliability and performance of the
  network?

Before transmission network planning can start, some basic activities have to take place to define operator's requirements and expectations. These include the following:

- Identify all the main nodes in the network (switch location, hub
  sites, collocated sites, and so on).
- Meet with customer(s), contractor(s), vendor(s), and/or partner(s),
  determine responsibilities for the transmission (leased lines, fiber,
  MW) network design and deployment, and complete the "scope and
  task delineation list" (who is doing what).
- Sign the nondisclosure agreements (NDAs) with all parties (customer, vendors, partners, and so forth) involved in the project.
- Identify potential microwave sites, MW link capacity requirements,
  and MW frequency bands/channels available and/or approved for
  the project and conforming to relevant ITU-R recommendations.
- Identify the available unlicensed microwave spectrum in case rapid
  deployment microwave systems (spread-spectrum) are required.
- Identify existing MW systems in the area and the source of information (microwave frequency coordination).
- Attain information (drawings, maps, and so forth) of the existing
  transmission facilities in the area (e.g., MW, fiber optics, copper) as
  well as PSTN offices and POPs of the local telco companies.

- Determine existing tower and other antenna mounting structures' capabilities, establish whether there is sufficient space for the MW radio equipment and antenna installation (provide site layouts and tower profiles), and verify access to those sites.[1]

- Find out all the customer-specific requirements (preferred equipment and services suppliers, power backup requirements, schedule, internal processes, and forth).

- Identify equipment and service resources (for international projects, try to find local companies).

- Develop preliminary transmission network build-out schedule.

The purpose of the *responsibility matrix* (sometimes also called the *scope and task delineation list*) will clearly state responsibilities related to all areas in the project. It is of great importance that all aspects are considered, especially when the customer also has to fulfill certain tasks.

It is important to notice that, in a large microwave project, like any other project, a compromise must be made between the speed of deployment, reliability of the system, and the price wireless operator has to pay for the network. Spending more money and time in the beginning and instituting a well executed planning process will guarantee a reliable network that will continue to work, even under unfavorable conditions. Implementing sound transmission engineering techniques such as ring topology, hardware redundancy when necessary, and so on will always prove to be good investment for the future.

4.2 Microwave Systems in Wireless Networks

4.2.1 Backhaul in Wireless Networks

Transmission is an important element in any wireless network, affecting both the services and service quality offered, as well as the costs to the wireless operator. Optimization of transmission solutions is thus certainly worthwhile from the business operator's point of view. In current wireless networks, transmission has been optimized for the narrowband circuit switched traffic, and this type of traffic will continue to dominate for some time.

The rapid growth in the number and diversity of the cell sites of *wireless service providers (WSPs)* has had a corresponding influence on the size and complexity of their transmission (and backhaul) networks. Since the introduction of cellular radio systems in the mid 1980s, there has been continued growth in microwave communications systems as a key component of the cellular backbone and access networks. In most cases, new wireless operators opted for use of microwave point-to-point systems for economic and deployment timing reasons.

In wireless networks, *backhaul* is defined as the portion of the network that carries the wireless calls from the cell site back to the *base station controller (BSC)* and *mobile switching center (MSC)*. Calls then are routed to the appropriate service termination points such as PSTN and/or the Internet. Once relegated as a second-tier priority in relation to RF and cell site deployment issues, wireless service providers (WSPs) today view their backhaul and transmission networks as strategic assets, since this portion of a network can run as high as 20 percent of an annual expense budget.

4.2.2   Wireless Network Design for Coverage and Capacity

Successful wireless network operators launch their services by providing adequate geographical coverage in key areas of the network, including major cities main roads and selected special tourist/business locations. Fast service rollout for a new network is of vital importance to enable cash inflow and investment payback.

Most wireless networks are initially designed for coverage, they have very low initial capacity requirements. Transmission and microwave engineers have to keep in mind that, very soon after the network launch, this strategy will change, and capacity requirements will increase dramatically in a very short period. The transmission network in such a launch phase is not usually optimized for link capacities or cost, but the transmission system and its management system should be selected for their ability to allow future network growth. During later phases, when the wireless network is redesigned for capacity, many new base stations will share the main transmission links of the initial launch network.

As the subscriber base increases, successful operators add new base stations, base station controllers (BSCs), and switches to serve all cities and improve roadside and indoor coverage. The number of macrocellular base stations can grow swiftly to hundreds and ultimately to several thousand, and the number of switches could grow to 20, 30, or more, resulting in a rapid increase in the number and capacity of transmission links. Cost optimization plays an increasing role during this phase. New services, mobile and fixed, have to be introduced with minimum rollout, with the same transmission infrastructure used for multiple services

As the competition in a wireless market grows, service differentiation and introduction of new value-added services play an increasing role. Operators who use their own transmission networks (microwave networks for example) tend to begin offering fixed services such LAN interconnect, voice, IP, and other services in any phase. As a cost-effective strategy, the some operators may use the same transmission network for both mobile and fixed services.

4.2.3   New Generation of Wireless Networks

4.2.3.1   3G wireless networks.   The central challenge for the telecommunications industry lies in creating a stable next-generation network platform that can support operators, service providers, and users efficiently and profitably for the foreseeable future. A significant portion of this challenge involves integrating the existing public switched network with packet-based technologies. Until recently, the business of telecommunications operators centred around providing basic voice services (i.e., dial tone) to a mass market and voice and data communications connectivity and services to large and medium-size businesses. Today, less than one percent of traffic in mobile wireless networks is data. The enormous growth experienced by today's mobile operators and service providers has been achieved predominantly in voice services. With the arrival of 3G, initially in the form of packet data capabilities for existing networks and subsequently by full 3G capabilities over new broadband wireless systems, this will change.

New technologies and technical solutions enable higher data volumes right now in existing networks. This development will continue with still higher bit rates over the air interface in the new 3G wideband code-division multiple access (WCDMA) and CDMA2000 based networks and 1xEV-DO (also called HDR). These increasing data traffic volumes mean that the share of the packet-based traffic in the total traffic mix in the mobile network is increasing while total traffic volumes are also rising rapidly. Evolution of the circuit switched networks into packet-based networks will take some time and will be done in well planned and managed steps so that the efficiency of the mobile network is preserved during the changeover phase. In many cases, basic mobile voice services are also growing quickly due to growth in the number of subscribers, which also contributes to the overall traffic increase and continues to require economic solutions for this type of traffic. Therefore, the well planned steps are vital to manage mobile operators' cash flows and to make full use of existing investments. It is in the interest of a mobile network operator to direct future transmission network strategy toward this expected increase in the penetration of advanced data services.[2]

Data traffic is inherently variable, and transporting it over the TDM network is inefficient. In a TDM-based approach (2G wireless networks), time slots are dedicated for connections regardless of whether information is actually being sent. In a *multiservice network,* the underlying network can be physically subdivided into multiple networks, one for each service (i.e., voice, data, private lines, and so on). Using an ATM-based infrastructure (3G wireless networks), much more efficient use of transmission network is possible, because it allocates bandwidth on demand based on immediate user needs.

However, as stated above, packet-based information over the mobile network will show rapid growth, and any reasonable network development plans have to take this into account and plan for a smooth and economic transition and an evolutionary path for the transmission network. Therefore, in broad terms, the transmission network must continue to provide well engineered and economically optimized solutions for the growing volumes of circuit based traffic while developing the strategy and readiness to cope with the even faster growing data traffic of the future. This type of transmission solution is needed in all parts of the wireless network, both in access networks with many points and low-capacity links and in core networks with high traffic volumes. This means for example that, in a wireless network, a transmission solution is needed that provides for efficient transport of many voice channels and that can evolve to also carry packet-based traffic, whether asynchronous transfer mode (ATM), Internet protocol (IP), or both.

The radio network will be connected to the core network by a backbone network (access and core transmission network), allowing wideband access and interconnection of subscribers. The 3G backbone network can use any transport technology but is certain to be based on packet technologies such as ATM and IP.

4.2.3.2 Deterministic and statistical multiplexing. In 2G wireless networks, *deterministic multiplexing* is applied whereby each connection is characterized by a constant bandwidth (e.g., one time slot). The minimum needed bandwidth over the physical link is then simply the sum of the constant bandwidths of the connections. Since the traffic characterization is not probabilistic, statistical gain is not available. 3G wireless networks use packet-switched (ATM) systems and statistical multiplexing. When several connections from variable-bit-rate sources are multiplexed together, a statistical multiplexing gain is obtained, because there is a certain probability that traffic bursts on different connections will not appear at the same time.

It is possible to maintain the same blocking probability with less bandwidth if statistical multiplexing is used instead of deterministic multiplexing. The price for it is that the *quality of service* (e.g., packet delay and loss) will not be ensured in a deterministic but in a probabilistic fashion.

Statistical multiplexing of data traffic can occur side by side with the transmission of the delay and loss-sensitive traffic such as voice and video.[3] Like voice telephony, ATM is fundamentally a connection-oriented telecommunication system. That means that a connection must be established between two points before data can be transferred between them. An ATM connection specifies the transmission path, allowing ATM cells to self-route through an ATM network. Being connection oriented also allows ATM to specify a guaranteed quality of service (QoS) for each connection.

4.2.3.3 Effect of ATM on microwave link planning.

In a process of dimensioning microwave point-to-point systems for ATM traffic, there are a number of issues to be considered. Since bit errors in the microwave system typically appear in multiples and spread less than the ATM header length, single-bit header correction feature may not improve cell loss rate as much as predicted and intended. The latest research shows that the bit error rate is degraded approximately one decade from the microwave radio system to the ATM CBR virtual circuit due to the cell loss. A general assumption based on the above could be to assume that any BER requirement should be one order of magnitude higher for ATM traffic.

In today's broadband networks (wireless or wireline), the traffic requirements will increase the total transmission capacity needs enormously. Many of today's 1E1/T1 links will be increased to STM-1/OC-3 and higher capacities and will require high-capacity SDH/SONET microwave radios. Aside from the capacity, these microwave radios need a very sophisticated error-correction technique to satisfy ATM transport layer requirements. Normally, in a fiber-optic system, BER should be 10^{-9} measured at the ATM CBR virtual circuit, and the same quality corresponds to BER = 10^{-10} in the microwave radio system.

Existing radio links were planned using parameters for PDH or SDH systems such as severely errored seconds (SES), errored seconds (ES), residual bit error rate (RBER), and background block error ratios (BBER) for which some time percentages of worst-month statistics have been allocated. When planning is based on 64 kbps ISDN-paths (ITU-T G.821), several different grades of quality can be applied, such as high grade, medium grade (four subclasses), or local grade. This applies mainly to existing PDH radio links. For mobile systems, one of the medium grade classes typically is applied (Class 3). When planning is based on primary level or above paths (ITU-T G.826), international portion and national portion are specified.

National portion has been subdivided into long-haul, short-haul, and access sections. This applies mainly to existing SDH radio links, while new international synchronous paths should be planned according to ITU-T G.828, which also applies to national and private synchronous paths. New ITU-T G.828 specifies recommended block-based error performance parameters for synchronous digital paths that may support circuit switched, packet switched, and leased circuit services. Synchronous digital paths meeting the objectives of G.828 will enable ATM traffic to meet B-ISDN-requirements of I.356.

Radio links planned according to G.826 using ITU-R F.1189 or F.1092 can fulfill ATM requirements if residual BER is lower than about 10 to 11 per 100 km path (in practice, four-level modulation or FEC fulfils this RBER requirement). If old PDH radio links that have been planned according to requirements of G.821 will be utilized, recalculations must be done for which fade margin corresponding to threshold level at about BER = 10^{-5} is needed.

If the frequency band is below about 17 GHz, multipath outage probability during worst-month must be calculated according to F.530-xx and the result compared to severely errored cell block ratio (SECBR) limit.

If the frequency band is above about 17 GHz, rain outage probability of worst-month must be calculated according to F.530-xx and the result compared to unavailability target applicable to the network (not yet specified to ATM, recommended spec. ITU-R F.1493).

Residual BER should be below 10^{-11}, which also can be measured by a suitable BER test.

4.2.4   Replacing Leased Lines with the Microwave System

4.2.4.1   About Leased Lines. The goal is to build a network that will provide reliable transmission facilities that are capable of delivering enough capacity for the present needs as well as to ensure seamless expansion in the future. Usually, transmission (transport) facilities are either leased or owned (copper, microwave, fiber optic), and sometimes they are a combination of leased and owned (usually microwave) facilities. In many cases, project managers and those involved in the time-to-market assessment make a conclusion that a faster way to build the network is to lease T1/E1 circuits from the local telephone companies rather than building their own microwave systems. That may not be the case in every situation and may prove, for a number of reasons, to be much more of an expensive and lengthy process than originally anticipated. A successful transmission platform includes necessary network functionality for the entire network life cycle, supporting network growth (scalability) and reconfiguration (flexibility) as well as fast multiservice deployment.

Dedicated service is a circuit between fixed end points, and the circuit is leased (or owned by an end user) from a common carrier. A dedicated line is obtained from the carrier on a monthly basis for unlimited use between the individual points to be connected. The service is available 24 hours a day, 7 days a week, 52 weeks a year, for exclusive use by that user. This service (usually T1/T3, E1/E3) may also be referred to as a *leased line, private line,* or *nailed-up* circuit.

In selecting a transmission medium for a network, service providers are faced with questions of quality, reliability, time to market, and cost. The temptation during the initial build-out stage is to utilize leased T1 or E1 service, because it avoids the capital outlay required for microwave networks, and it is generally readily available. Once in operation, a maintenance problem is simply addressed by placing a phone call to the carrier to report the problem. However, the decision to use leased facilities may also mean that the service provider sacrifices reliability, control, ownership, and return on investment.

It is very important to notice that, even when the transmission network is completely leased, it still requires a great deal of engineering effort from the wireless operator's side. Engineering cannot be left to the carrier(s) to make decisions on the leased circuits' routing, installation, project management, and testing without any supervision from the wireless operator. One of leading causes of delay in getting T1/E1 lines installed is the "unprepared equipment room." It is important to coordinate with carriers and cell-site construction team requirements for the equipment room/shelter. This could include ducts or cable ways for the lines, AC and/or DC power, adequate space, heating and air-conditioning, solid grounding, and a very common requirement by carriers for a plywood board on the wall for terminal blocks, NIUs, punch-down blocks, and so forth.

In addition, for any maintenance problems on the leased line, the time to repair is at the mercy of the carrier. Therefore, it is important to establish a good working relationship with the facility (Telco) provider so as to understand the organization of the company and to access key departments when help is needed.

4.2.4.2 Advantages of microwave radios. The question is whether to build and own a transmission network or to lease facilities (lines) from existing carriers and/or operators. The answer will be different in different situations. It is important to realize that transmission network planning, design, and implementation will directly affect all stages of the wireless network build-out. Introduction of 3G wireless networks with the increased capacity requirements and packet data architecture will also have a great impact on the future transmission network design and deployment.

Owned facilities in any network usually involve planning, building, and maintaining the microwave network. Some of the advantages of microwave radios are listed below:

- MW system meets superior reliability, higher security, and more demanding performance and quality standards.

- One can reuse existing infrastructure such as shelters, towers, and monopoles to set up new microwave radio systems.

- The user has total control over the site access and restoration time.

- Expansion and future relocation are easy.

- Microwave systems do not require right-of-way permits from local governments as is necessary for buried wireline systems.

- MW radio has an operational life (>15 years) long after the leased line payback time (2 to 4 years).

Some disadvantages of microwave networks are as follows:

- It involves, high capital investment on day one (unless the vendor is financing it).

- Microwave system design is required, including frequency coordination and spectrum licensing (which may or may not be a big issue).

- Microwave network operations and maintenance are required.

The last statement needs some additional explanation. In a well designed and properly implemented microwave network, there is very little required maintenance. Assuming that high-quality microwave equipment from a reputable vendor is used, long life expectancy and trouble-free operation is almost guaranteed.

4.2.4.3 The economic model and the payback period. Transmission is an important element in any wireless/mobile network, affecting both the services and service quality offered as well as the cost to the mobile operator. Optimization of transmission solutions is thus certainly worthwhile from the business operator's point of view. Microwave access, based on point-to-point microwave radios, is the dominant technology in base-station access networks. It offers the fastest means for network rollout and capacity expansion. When using microwave radio transmission, an operator saves on operational expenses as compared to laying his own cable or leasing connections. At least two-thirds of all base station connections worldwide are based on microwave radios.

The main problem with leased lines is the high recurring cost that will always exist unless the intention is to replace it with a privately owned system. In addition to the recurring cost, there is a typical service charge, usually several thousand dollars per facility, and a construction charge, from several thousand to hundreds of thousands, if the carrier is required to build facilities to a site. Leased line service costs also include a one-time hookup (service) charge, a monthly lease rate, and, in some regions, a monthly inter-LATA or long distance carrier charge. The cost of nonprotected, single T1 service can vary widely by geographic area from a few hundred dollars to over $2,500 for the hookup charge, from under $250/month to over $700/month for lease charges, and up to $15/km/month if distance charges are involved.

Another problem with leased facilities is their limited capacity and the fact that leased transmission facility cost is linear with respect to bandwidth. For example, if twice the capacity is desired, then twice the facility must be leased, thereby doubling the cost. Leased facilities are also inflexible in network design, and it is a lengthy and costly process to reconfigure leased lines to address ongoing changes in the network. Protected (i.e., redundant) T1 service is usually not offered in the standard tariff, and it normally results in monthly leasing charges that can approach twice that of single T1 service.

Every digital network project is different and requires a detailed "business case" type analysis of "owned versus leased" transmission facilities. While a network is relatively small, it will generally be more cost effective to employ leased transmission lines (assuming, of course, that they are available) due to the ability to closely match supply to demand. However, at some point, which depends on many factors such as cost of leased transmission versus cost of building infrastructure, it will become more cost effective to build and own the infrastructure in some or all parts of the transmission network. Most operators require a payback time of 24 months to justify investment in a new network technology.

Let us assume that the 4T1 microwave system cost, fully engineered, installed, and tested, is $50,000 per hop (example shown in Table 4.1). Not only is the total cost after four years still the same, it will also provide system expansion for additional three T1s, thus resulting in the cost per T1 actually being only $12,500. On the other hand, leased T1 has recurring costs of $400.00 per month (typical in the U.S.A.), and after four years the cost for the leased T1 would grow to $20,200—almost three times that of microwave T1 circuits.

In many countries where leased lines are not readily available or cost a lot more, it would be even easier to prove the advantage of building a microwave network. If not one but multiple T1/E1 circuits are required to connect two points, the superiority of the microwave system and its quick break-even point (payback time) are usually very easy to demonstrate. Equipment supplier financing is an attractive

TABLE 4.1   Comparison of Leased Lines versus Microwave Systems

Expenses	4T1 microwave system	Leased 1T1
Up-front charges	$30,000 (turnkey)	$1,000
Monthly charges	$ 0	$400
After 12 months	$30,000	$5,800
After 12 months per T1	$ 7,500	$5,800
After 48 months	$30,000	$20,200
after 48 months per T1	$ 7,500	$20,200

means of acquiring microwave systems, particularly for infrastructure build-out requirements. In general, equipment supplier leasing is the financing method preferred by users.

For each of the technologies, it is assumed that there will be ongoing maintenance and operational costs. The maintenance costs are usually assumed to be around five percent of the equipment cost on an annual basis, and the operational costs are assumed to be two percent, but it is recommended to use some real data (if available) from a similar type of network, in the same country, and using the same type of equipment.

4.3 Microwave Systems in Utility Telecom Networks

4.3.1 SCADA

Microwave radios systems have been used in many utility companies since the early 1950s. Electric power and other utility telecommunications networks have some specific requirements that are different from those of other telecom networks, such as (among others) supervisory control and data acquisition (SCADA), *powerline fault location, loop and ring protective schemes,* and *direct transfer trip protective relaying.* SCADA systems are used extensively by power, water, gas, and other utility companies to monitor and manage distribution facilities.

Direct transfer trip transfers local protective relay tripping signals hundreds of miles to operate a distant circuit breaker. The allocated time to clear a high-voltage fault is typically three cycles, meaning about 50 ms at 60 Hz, and the large part of the sensing and operation times are specific to devices such as protective relays and circuit breakers whose characteristics cannot be changed. The only variable that can be controlled is the communication channel transit time, which includes free-space delays, channel banks, microwave terminals and repeaters, waveguides, and so on. Typical transit (channel delay) time objective for the communication channel is 10 to 15 ms end to end.

It is important to notice that transmission and therefore microwave systems for wireless networks, telephone operators, and government or utility companies will have very different sets of requirements to be considered.

4.3.2 Electric Transmission Towers in Telecom Networks

Large electric transmission towers provide a corridor between generation stations and substations, placed in remote, out-of-the-way locations and run through less-expensive territories. Typical voltages on these transmission towers range from 138,000 to 500,000 V. Their location and 150+ ft height allow for placement of larger antenna arrays and excellent antenna elevation. These towers are ideal as antenna locations for propagating wireless signals over large areas. Precise engineering and extreme care must be used when placing RF and/or microwave antennas near transmission conductors to avoid the danger of electric arcing.

From a strictly microwave prospective, these towers and power lines are not an obstacle to installing a microwave system, even if they obstruct the LOS of the microwave system. Signal loss due to the obstructed Fresnel's zone would be only 1 to 2 dB, depending of the type of the tower and its construction.

Installation of the microwave systems using electrical utility poles and towers must be carefully examined, since they may not fulfill the twist and sway requirements of the microwave antenna mounting structure—especially for higher frequencies.

4.4 Topology and Capacity Planning

4.4.1 Transmission Network Capacity Requirements

Aside from immediate capacity requirements, the likely future traffic capacity of nodes connected in the transmission network should be considered so as to appropriately dimension the transmission links. Operators usually dictate the transmission network utilization factor. The utilization factor tells us how much of the total capacity of the certain link is used on day one and how much is a growth margin reserved for future expansion. In a microwave network, the utilization factor should not be more than 70 to 80 percent, leaving at least 20 to

30 percent for network capacity upgrades. This rule of thumb may need modifications in the case of long chains with many microwave links connected in tandem and/or other operator-specific requirements.

In addition, the total number of links in a chain or ring of nodes should be such that the transmission links between the included nodes can be relatively easily and inexpensively upgraded (e.g., by simply replacing a modem unit or via a simple software change in a microwave link) to accommodate increased node traffic. This expansion should be accommodated within the chosen topology, without the need for major rerouting of transmission paths.

For a small, low-capacity network, a PDH solution is usually sufficient whereas, for the higher-capacity systems, an SDH network is a preferred solution.[4]

The *network traffic plan* is a document reflecting all these requirements in terms of *T1/E1 plan* as well as the *channel plan*. A channeling plan describes the routing of individual channels between microwave sites in their respective T1/E1 trunks and is therefore based on the T1/E1 plan. A T1/E1 plan is usually a drawing that shows the traffic flow of the system in terms of T1 or E1 circuits.

4.4.2 Chain and Tandem Topology

The overall transmission performance of a tandem network is largely influenced by the propagation characteristics of the individual hops. It is sometimes possible to achieve the same overall physical connection by using different combinations of hop lengths. Increasing the length of individual hops inevitably results in an increase in the probability of outage for those hops. On the other hand, such an approach could mean that fewer hops might be required, and the overall performance of the tandem network might not be impaired.

In the wireless network, this type of configuration consists of linking RBS sites in a chain such that every RBS site in the chain acts as an active repeater for the previous one (see Fig. 4.1). This figure illustrates two chains converging to a common BSC and, in this particular case, the configuration can also be considered as a "tree." A common application of chain is the connection of RBS sites along roads (called *highway cell sites*). Closer to the BSC, where the capacity is higher, it is recommended to have some degree of protection (1+1 configuration).

4.4.3 Simple Star Network

Figure 4.2 illustrates a common pattern in which all RBS sites are directly connected to the BSC to form a star network. The advantage of

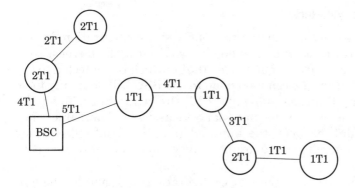

Figure 4.1 Chain/tandem network configuration.

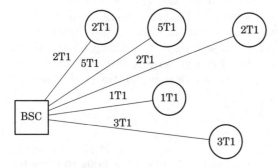

Figure 4.2 Simple star network.

this configuration is that the RBS sites may be established to expand capacity requirements in a particular area separately from capacity requirements in other parts of the network, and the network may be gradually taken into service in accordance with the establishment of new sites.

This configuration also has the following disadvantages:

- It involves a large number of incoming BSC routes and their corresponding antennas. This may cause space and strength problems for antenna support structures. Robust structures are generally more expensive.

- The high number of incoming routes may lead to problems in finding a sufficient number of available channels.

- Some sites may be situated too far from the BSC, thus increasing fading probabilities. Therefore, this configuration is used mainly in leased lines networks.

4.4.4   Network with Hubs

Figure 4.3 illustrates another option of the star configuration. In this specific case, the connection is made in two stages. The farthest sites are connected first to a common node (hub), which is connected to the BSC. The link from the common node to the BSC will generally have higher capacity than the individual RBS site connections. To handle longer distance, it may be necessary to assign a lower-frequency band to the link between the hub and the BSC. Higher-frequency bands are therefore reserved for the connection of the individual RBS sites.

The main drawback to the star configurations is generally the vulnerability for hardware failure in the common node—the hub.

This type of configuration is suitable for the next generation of wireless networks utilizing statistical multiplexing. For example, if the hub site requires 5T1s for connection to BSC in a 2G network with deterministic multiplexing, we can assume that not all the sites in a 3G network and connected to the hub will be fully utilized all the time. Therefore, there is a chance that we may require only 4T1s between the hub and BSC and save on the transmission facilities, either leased lines or microwave.

4.4.5   Ring Topology

By using radio links in a ring topology network, each node in the ring (i.e., each base station in wireless network) is provided with two alternative routes (see Fig. 4.4). In the event of a failure in one link, the traffic can be sent in the other direction of the ring. The main advantage of this configuration is that it improves the availability of the network and can be built using PDH as well as SDH technology. If the

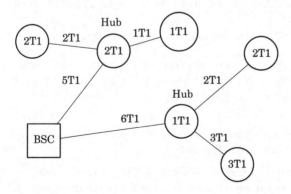

Figure 4.3 Star network with hubs.

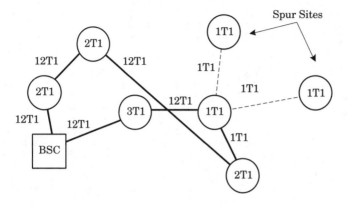

Figure 4.4 Ring (loop) topology.

ring has sufficient capacity to carry all the traffic from every site in both directions, then complete redundancy has been achieved.

The capacity requirement is the total sum of the individual capacity requirements. A PDH ring with maximum capacity of 34 Mbps can then normally handle 16 nodes, with an average capacity of 2 Mbps. Note that the traffic generated at each node, along with grooming, may result in other capacity requirements.

Unavailability time caused by hardware failure is reduced without the necessity of doubling the radio equipment. That means that an unprotected (1+0) configuration can be used for all the links forming the ring without sacrificing the availability of the network. Most links in the ring use a higher capacity than would be used in a simple tandem chain. This means that each link works with lower system gain than in a corresponding tandem chain, which is compensated by less fade margin needed due to the ring protection. As a result, the links in a ring-protected network should be able to use smaller antennas.

It is important to notice that the physical layout does not necessarily have to form a ring; it is the actual flow of traffic that determines the ring topology.

4.4.6 Mesh Topology

The mesh topology is a mixture of the previously described configurations and is currently employed to improve the availability to the network (see Fig. 4.5). This configuration is not really a cost-effective solution, so it is somewhat rare. Furthermore, the traffic distribution presents more complexity in the physical layer. Other configurations normally exhibit equivalent reliability at lower cost.

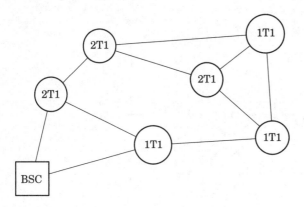

Figure 4.5 Mesh topology.

4.4.7 Transmission Network Optimization

The main purpose for having digital *access cross-connects (DACS)* in the transmission network is *grooming functionality*. This way, operator can efficiently "fill" or "groom" DS0s into the T1/E1 circuit and thereby optimize the use of the network links. When the physical capacity on the radio hop is not fully utilized by the existing service, it might be a better solution to install a DACS and add new traffic over the same T1/E1 circuit rather than upgrading the radio. When upgrading the radio, a new frequency plan, new link budget, and new interference calculations have to be made. In addition, for high-priority circuits, the digital access cross-connect can automatically reroute the traffic in a ring/meshed network if a fault in the primary path occurs.

A word of caution: Although efficient circuit utilization is a good thing, and many telecom and wireless operators prefer this solution to network upgrades, it may not leave room for future changes. In other word, for example, if original voice services required 15 DS0s, and the other 9 DS0s were left for future expansion, it may not be a good idea to use them for something else (for example, to overlay data). If we do so, an increase in voice traffic may create a problem. Overlaying a new network over the existing network without adding any new transmission capacity may prove to be a dangerous thing. Perhaps building a new network now can save lots of time, effort, and money later. For detailed information on planning transmission networks see Ref. 3.

4.5 References

1. Timiri, S., "RF Interference Analysis for Collocated Systems," *Microwave Journal,* January 1997.

2. Bates, J., *Optimizing Voice in ATM/IP Mobile Networks,* New York: McGraw-Hill, 2002.
3. Lehpamer, H., *Transmission Systems Design Handbook for Wireless Networks,* Norwood, MA: Artech House, 2002.
4. ITU-T G.803, "Architecture of Transport Networks Based on the Synchronous Digital Hierarchy (SDH)," 03/2000.

Chapter

5

Microwave Network Design

5.1 Introduction

After the preliminary microwave network plan has been approved, detailed microwave network design has to be completed. Site acquisition, microwave network design, RF design (in case of wireless network build-out), and interference analysis are done simultaneously. In most cases, the results are mutually dependant. That means that none of these activities can be done without consultations with and input from the other three. It also means that a project manager has to make sure that these groups of experts talk to each other on a daily basis, which can sometimes present a challenge. The results and deliverables of the microwave network design process will be used during the deployment stage for the actual installation and testing of the microwave system. Microwave path (link) calculations are performed as a part of detailed microwave system design, and all the detailed hardware requirements *(bill of materials)* are defined based on this information. The microwave design software tools are used for detailed path engineering and interference analysis.

5.2 Spectrum Management

5.2.1 Availability of Spectrum

Congestion of the radio frequency spectrum requires sharing many frequency bands among different radio services and among the different operators of similar radio services. National administrations will allocate some or all these bands for fixed microwave radio use in line with local requirements. To ensure the satisfactory coexistence of the systems involved, it is important to be able to predict, with reasonable accuracy, the interference potential among them, using prediction procedures and models that are acceptable to all parties concerned and that demonstrated accuracy and reliability.

Before microwave network planning commences, the operator must determine the available frequency bands and channel plans that are specific to the country in which the network will operate. There has been a recent trend toward spread spectrum microwave links that do not need to be individually licensed. This includes Part 15 transmitters operating in several industrial, scientific, and medical (ISM) bands, the 5 GHz unlicensed national information infrastructure (U-NII) bands, and many new bands that are licensed by geographical area.

Users of a common band of radio frequencies must follow a procedure of radio frequency coordination so as to minimize and control potential interference among systems. Frequency coordination is a multilateral process that involves the cooperative sharing of technical operating information among parties utilizing the same spectrum. The procedures are based on the Federal Communications Commission's (FCC's) coordination and licensing requirements (found in Rule Part 101) as well as on related industry practices that have evolved over the years.

The radio license applicant must determine if the planned radio system will experience any interference from the existing environment, and vice versa. Potential interference can be calculated for three different cases:

1. Interference between microwave stations

2. Interference between microwave stations and Earth stations

3. Interference between microwave stations and a geostationary satellite in orbit

Regulations for telecommunications are contained in Title 47 of the U.S. Code of Federal Regulations (otherwise known as the FCC Rules), and rules for the use of microwave transmitters in the bands above 3 GHz for common carriers are contained in Part 101. Part 101 consolidates the old Part 21 and Part 94 rules for the bands above 3 GHz into one set of rules for both common carriers and private operational fixed users. All frequency bands under Part 101 are available for both types of user.

The FCC does not maintain an on-line copy of the rules; however, the Government Printing Office (GPO) does have an on-line search location at the following web-page: http://www.gpoaccess.gov/cfr/index.html. To find the Part 101 rules, input "47CFR101" as the search criterion.

5.2.2   Intersystem and Intrasystem Frequency Coordination

Sometimes, an operator may be able to obtain a number of frequency allocations as a "block," enabling network planning to be performed in advance, without the risk of interference from other users. Most regulatory authorities also operate a local link length policy whereby the length of a particular path will determine what frequency bands are available from which the operator may choose. Typically, the shorter the path, the higher the frequency required. The local requirement for equipment type approval will also vary from country to country, ranging from a simple paperwork exercise to a full product test program to local standards. Type approval is generally the responsibility of the radio supplier, but an operator should ensure that all requirements are satisfied before any links are deployed.

The first step is to perform *intrasystem frequency coordination* (within its own network) and then, if the results are satisfactory, perform *intersystem frequency coordination*. A radio license applicant must determine if the planned radio system will experience any interference from the existing environment or create interference within it. Potential interference can be calculated for the three different cases described previously.

The design of radio links to achieve a particular performance objective is based on equipment and propagation behavior, taking account of intra- and intersystem interference. Many times during the inter-system interference analysis, it may become necessary to change certain parameters of the microwave link and therefore modify the original microwave transmission design. Intersystem frequency coordination includes a detailed frequency search to identify available frequencies for a proposed microwave path based on provided parameters. The study includes a search of all combinations of frequencies and polarizations. Alternative parameters, such as antenna or equipment changes, are studied to maximize frequency availability.

Interference analysis (including simulation) of specific interference situations involving space and/or terrestrial systems, including the identification of possible interference mitigation techniques, is done at this stage of the microwave network design. If necessary, frequency coordination across the border with other countries *(transborder coordination)* is also performed.

5.2.3   Spectrum Sweep

The key aspect of the frequency coordination procedure involves informed radio frequency planning. Radio systems should be designed such that they do not to cause or suffer objectionable interference while operating with other existing or planned systems using the same frequency band. Sharing coordination data among users facilitates this coordination so that accurate and up-to-date information is available with which estimates of potential interference can be made during the system design stage. Radio frequency interference studies and frequency coordination are necessary not only when designing a new system but also when on is assessing the potential interference effects of other users' radio construction proposals on existing and planned systems. Thus, coordination is required when one party initiates construction plans as well as when reacting to other parties' plans. The results of these studies will indicate whether there is potential interference and whether redesign or relocation of the planned MW system is required.

In many cases, the most reliable information about the potential interference cannot be generated by calculation, since there may be little or no information about existing terrestrial or satellite systems in the area. (This is often a case outside North America and Europe.) The best way is to sweep the entire spectrum using test equipment at the future microwave-system antenna location (at the antenna centerline height) and determine the interference potential at that location. Sweeping the frequency spectrum at the ground level, although a much simpler and cheaper solution, will not produce accurate results.

The primary tool used for accomplishing the task of interference analysis (spectrum sweep at the microwave site) is the spectrum analyzer, which shows power level as a function of frequency. The result is a spectrum analyzer plot showing all potential interference in the applicable band.

5.3 Interference Effects and Frequency Sharing

5.3.1 Interference Paths

Interference is the general term for any kind of radiation disturbance on radio-link systems. In this text, however, only interference caused by radiation from radio systems will be considered. The government requires users of the radio spectrum to frequency coordinate their planned and existing point-to-point microwave radio systems with other users of the radio frequency spectrum.[1] Such coordination is a prerequisite for any microwave radio license application submitted by a microwave radio system operator.

The license applicant must determine if the planned point-to-point radio system will experience or cause any interference within the existing environment. The results of this calculation will indicate whether there is potential interference and whether a redesign or relocation of the planned MW system is required. In addition, many countries place some very specific requirements on the MW equipment that may be installed. Channel plans, maximum transmit power at the antenna port, and channel separation requirements can differ from country to country.

Considering the case of one transmitter and one receiver (which may be collocated), interference may propagate via the following paths (see Fig. 5.1):

1. From equipment housing one unit to that of another unit, between components housed in the same cabinet, or among units in the same telecommunications room

2. From the transmitter antenna to the receiver's equipment housing

3. From the transmitter's antenna to the receiver's antenna

4. From the transmitter's equipment housing to the receiver's antenna

5. As spurious signals in the power supply system

It is assumed that, if one follows local rules and regulations and performs appropriate installation procedures, interference paths 1, 2, 4, and 5 will be eliminated. On that assumption, only interference between antenna (case 3) systems must be considered.

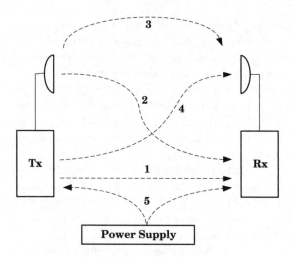

Figure 5.1 Interference paths.

5.3.2   Collocation of Radio Stations

Collocation is a general concept that refers to multistation sites consisting of numerous transmitters and receivers installed within a limited geographical area. The site often consists of a number of antennas that are all mounted on the same mast or distributed among a small number of closely positioned masts.

Collocation is a logical and creative siting strategy, and it can be approached in two ways. The first approach is for all operators to try to negotiate collocation of their RF and/or microwave equipment on a common tower. The second would be to outsource the business of antenna installation and rent tower space from independent companies. Such companies usually offer site engineering, acquisition, and installations services, and they handle routine maintenance.

Based on the latest FCC requirements, wireless service providers in the U.S.A. will have to prove that they meet safety guidelines for all cell sites constructed, licensed, and activated before October 15/1997. They must ensure that cell sites comply with safety limits for human exposure to radio frequency emissions. Some of the applicable requirements for existing and/or new operators are as follows:

- RF emissions of all new cell sites still must be assessed and documented before the facilities are activated.

- Anytime a licensee renews its operating license, it must document the compliance of all sites.

- Anytime an operator modifies a site in any way, it must prove that the site remains compliant.

The collocation trend in the industry actually can create compliance challenges that operators otherwise would not have encountered. The reason is they must submit compliance records for their own equipment and for the equipment owned by collocation tenants at the site. This is particularly important for rooftops where multiple operators install transmitting facilities. In certain situations, and depending on the site accessibility to the public, if emissions exceed the maximum allowed exposure levels, any company that contributes five percent or more to the RF emissions in that area is responsible for mitigating the problem. If the tower meets certain height criteria or if the site operates at low power levels, the operator could be exempt from such routine procedures, but it must be proven that the site falls within this category.

Three different compliance procedures are acceptable. One approach relies on paper studies to calculate exposure levels based on the type of equipment and operating conditions at the site. Another approach uses industry accepted software tools that employ computer modeling and simulation techniques to perform the calculations. A third method is to take actual measurements at the site location. RF interference is a problem not only in collocated systems; it can occur in any RF system that can interact with existing systems. It is simply that collocation additionally complicates interference analysis and control.

Microwave equipment has to be included in all calculations aimed at determining interference levels between operators, including safety and RF exposure issues. The main purpose of such a study is to ensure that the installation of the new microwave link will not substantially degrade performance of existing microwave links at and near that location. The applicant company, according to its interference/availability criteria, must determine the allowed degradation that will be caused by the new microwave link. If, for example, the applicant company has an objective of 0.005 percent of unavailability per annum per link (due to propagation only—mainly rain), this objective cannot be exceeded by the sharing addition. From the point of view of the applicant, the degradation caused by the new link must allow all of the individual unavailabilities of all other links to remain below the desired value (0.005 percent per annum per link, in this example). This is a primary condition for the approval of the sharing, but not the only one; infrastructure and legal questions must always be considered as well.

5.3.3 Minimizing Near and Far Interference

The term *near interference* refers to interference contributions arising from transmitters and receivers situated at the *same site (collocation)* or in its immediate vicinity. Other interference contributions are termed *far interference*.

Near interference (also known as *on-site interference*) means that a transmitter in one site interferes with a receiver in the same site. *Intermodulation* is a typical form of near interference disturbance. It occurs because of different kinds of nonlinear processes taking place in the equipment that forms the transmitter and receiver.

An intermodulated signal is formed by the addition of interference signals and their integer products. Intermodulation disturbances are generally not expected to affect radio links using waveguides and parabolic antennas because of the higher degree of antenna isolation for typical radio-link antennas. In addition, the intermodulation products and frequencies of radio systems operating in other frequency bands usually fall outside the frequency bands used in microwave communications. By allocating the same duplex band (lower/upper) to all the transmitters at the same site, all receivers at the site will automatically operate in the other part of the duplex bands (upper/lower), and near interference, in most cases, can be neglected.

Another important characteristic that should be considered when calculating the effect of near interference is the coupling loss between two antennas located at the same site (see Fig. 5.2). This is a very important issue when collocating new with existing microwave equipment, and the coupling loss between two antennas should be approximately 80 dB. In reality, this will depend on the distance and angles between the two antennas.

Far interference (also known as *far-field interference*) is present when a received signal is disturbed by signals that are sent on the same channel *(co-channel interference)* or an adjacent channel *(adjacent-channel interference)* and are generated by a transmitter located far away from the receiver (Fig. 5.3). The influence of far interference

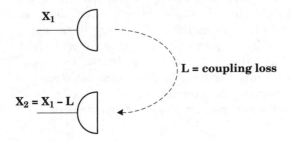

Figure 5.2 Coupling losses between antennas.

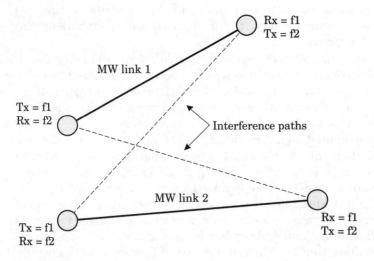

Figure 5.3 Far interference.

is first noticeable during fading conditions as a deterioration of the receiver threshold level; that is, as a decrease of the path's fade margin. In interference-free reception, the path fade margin is solely dependent on the path parameters. Far interference is often the primary factor that limits the number of paths that can be set up within a given geographical area. Planning an interference-free (in this case, far interference) network will involve the following considerations:

- Knowledge of the geographic locations of the sites, the layout, and dimensioning of the radio-link paths

- Equipment data

- Existing network frequency assignments

- Reasonably accurate radio-wave propagation models

During reception with interference, the fade margin changes, because the receiver's threshold is degraded—assuming that the bit-error ratio is kept unchanged. The degradation is generally the result of two contributions: the resulting (total) interference level at the receiver and the receiver noise level. The most serious problem caused by interfering transmitters occurs when they transmit at the frequency to which the disturbed receiver is tuned, producing *co-channel interference*. In some rare cases, serious disturbances may arise even when the interfering signal lies in an adjacent and separate channel rather than the channel containing the desired signal *(adjacent channel interference)*, but this is normally a minor problem in microwave point-to-point networks.

The ability of digital channels to tolerate interference depends on the modulation scheme.[2] In particular, modulation scheme that requires low C/I for a certain bit-error ratio is more tolerant of interference. Robust modulation schemes are, for example, 2PSK and 4PSK, whereas more complex modulation schemes such as 128QAM require much larger C/I-values.

5.3.4 Frequency Planning

5.3.4.1 Frequency planning objectives.
The objective of frequency planning is to assign frequencies to a network using as few frequencies as possible and in a manner such that the quality and availability of the radio-link path is minimally affected by interference. The following aspects are the basic considerations involved in the assignment of radio frequencies:

- Determining a frequency band that is suitable for the specific link (path length, site location, terrain topography, and atmospheric effects)

- Prevention of mutual interference such as interference among radio frequency channels in the actual path, interference to and from other radio paths, interference to and from satellite communication systems, and so on

- Correct selection of a frequency band that allows the required transmission capacity while efficiently utilizing the available radio frequency spectrum

Allocation (of a frequency band) refers to the frequency administration of a frequency band for the purpose of its use by one or more services. This task is normally performed by the ITU. *Allotment* (of a radio frequency or radio frequency channel) is the frequency administration of required frequency channels of an agreed frequency plan adopted by a competent conference. These frequency channels are to be used by one or more administrations in one or more countries or geographic regions. *Assignment* (of a radio frequency or radio frequency channel) is the authorization given by an administration for a radio station to use a radio frequency or radio frequency channel under specified conditions.

Allotment and assignment are created in accordance with the Series F Recommendations given by the ITU-R. The allotment consists of one or more alternative radio frequency channel arrangements.

These arrangements are used in accordance with the rules of the local administration in a country or geographical region. In most applications, however, frequency bands and frequency channels are already selected and provided to operators.

5.3.4.2 Frequency channel arrangements. *Channels* are segments (subdivisions) of a frequency range or a portion (frequency band) of the electromagnetic spectrum. Every channel has a specified bandwidth and, depending on the capacity of the link, a certain number of carriers can be accommodated in the band. For instance, a frequency raster may include four adjacent 28-MHz channels (applicable for 34-Mbps links), but each of these channels can be further divided in four 7-MHz channels (applicable for 8 Mbps). To enable four 7-MHz channels to be included within one 28-MHz channel, the center frequencies of the 28- and the 7-MHz channels cannot coincide. Likewise, each 7-MHz channel may be divided in two 3.5-MHz channels (applicable for 2 or 2 × 2 Mbps).

The available frequency band is subdivided into two equal halves: a lower (go) and an upper (return) duplex half. The frequency separation between the lowest frequency in the lower half and that of the upper half is known as the *duplex spacing* (see Fig. 5.4).

The duplex spacing is always sufficiently large that the intended radio equipment can operate interference-free under duplex operation—meaning that one channel has one frequency for transmitting and one for receiving. The width of each channel depends on the capacity of the radio link and the type of modulation used. The ITU-R recommends frequency channel arrangements according to homogeneous patterns given as follows:

- Alternated
- Co-channel band reuse
- Interleaved band reuse

Interleaved channels refers to an arrangement of radio channels in a radio link in which additional channels are inserted between the principal channels. The center frequencies of the additional channels are shifted by a specified value, which is a significant proportion (such as a half) of the channel bandwidth from the center frequencies of the principal channels (see also Recommendation ITU R F.746).

5.3.4.3 The frequency planning process. Channel frequencies may be available on a *link-by-link basis* or as a *channel block* and may be freely used by the operator. In the first case, it is common that a local frequency administration coordinates the use of the frequencies among different users. In the second case, it is up to the operator to coordinate the use of the channels within its own network. Local frequency administrations usually keep track of the use of available frequency bands and the corresponding channel distribution. Several operators may be forced to share the same frequency band but different channels, thus making it necessary to control such radio-link parameters as transmitted power, site coordinates, antenna heights, and so on.

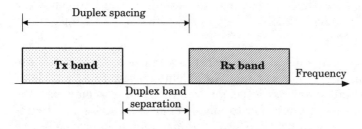

Figure 5.4 Frequency band subdivision.

The most important goal of frequency planning is to allocate available channels to the different links in the network without exceeding the quality and availability objectives of the individual links because of radio interference. Frequency planning of a few paths can be carried out manually but, for larger networks, it is highly recommended that one employ a software transmission design tool.

The frequency planning process can be described as follows:

1. Define the overall structure of the network by determining the location of all the nodes that have to be connected.

2. Allocate the appropriate quality and availability objectives for every portion of the network (no frequencies are involved in this step) and perform quality and availability calculations.

3. Estimate the traffic requirements and capacity. It is a good practice to start frequency planning with highest-capacity links in the most concentrated node. This will normally result in the number of frequencies needed in the network, and other links should reuse the same frequencies. In some cases, it may be necessary to use channels from more than one frequency band as a result of the limited number of available channels in the first selected band.

4. Start assigning a duplex half (lower/upper) for the transmitter in the sites of the network. Generally, near interference should be avoided as much as possible by strictly allocating the same upper or lower duplex half to all transmitters (or receivers) on the same site. Generally, two alternatives are possible:

   – In a chain of sites, there will be alternating lower/upper sites; that is, the transmitter in site 1 is L (lower), site 2 is U (upper), and site 3 is L, and so on.

   – A microwave ring should always have an even number of hops. In a ring with an odd number of sites, the transmitter of the first site will be assigned the same duplex half as the receiver of the last site (which is the first site in a closed ring), causing serious interference.

5. Consider antenna discrimination aspects in the early stages of frequency planning. For instance, in a common site (e.g., a node or hub), the links having sufficient separation angle may use the same channels. In addition to angle separation, distance separation (coupling loss) between two antennas may also provide a certain degree of discrimination.

At microwave frequencies, antenna discrimination increases rapidly with angle separation and is an extremely efficient factor in suppressing interference. Thus, if the two links are not closely aligned to a common line and with the (upper or lower) transmitters transmitting in the same direction (in other words, no *overshoot*), it is normally possible to reuse frequencies between two such links.

6. Use antennas that have high front-to-back ratios and large sidelobe suppression. These result in good frequency economy and, in the final analysis, good overall network economy. High-performance antennas may be a suitable alternative.

7. Reuse frequencies and polarization as often as possible.

8. Perform a new quality and availability calculation (after the frequency allocation) and identify links that do not meet the quality and availability objectives. Far interference calculation is performed, and the receivers having relatively high threshold degradation values are probably a part of the links that do not meet the quality and availability objectives.

    Make the appropriate changes (polarization, channel, frequency band, antenna size, and so forth) and ensure that a new interference calculation gives lower threshold degradation values.

9. In some situations, higher output power of a transmitter may improve the quality and availability figures without a significant interference contribution to the network. These favorable situations, however, are not very common, so it is not advisable to use a higher output transmit power than necessary. It is a good idea to start frequency planning with the lowest available output power. Then, if the choice is between higher transmitter output power and larger antennas, choose (if possible) a larger antenna.

10. Repeat step 8 until the quality and availability objectives of all portions of the network are accomplished.

5.3.4.4 Network topology and frequency planning. Interference aspects may severely limit the number of links in a network if appropriate caution is not exercised in the earlier stages of frequency planning (see Fig. 5.5). In what follows, some general aspects, based on former sections, are illustrated.

Frequency planning is similar for both *chain* and *cascade configurations*. Since paths of a chain have very sharp angles, using the same channels by changing polarization (H/V) may be a good alternative to using two alternate channels in the chain. Figure 5.5 shows the same channel used alternately with horizontal (H) and vertical (V) polariza-

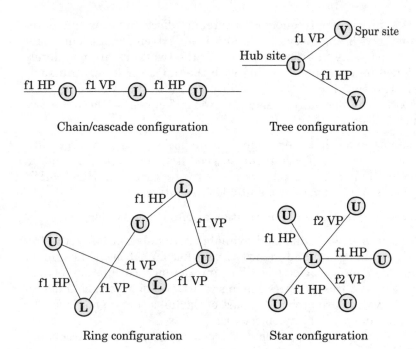

Figure 5.5 Frequency planning for different network topologies.

tion. Upper (U) and lower (L) duplex halves for the transmitters are illustrated in each site.

In the *tree configuration,* and for sharp angles, polarization discrimination ensures the possibility of using the same channel with different polarization (HP and VP). Both transmitters on the common node have the same duplex half (U).

In the *ring configuration,* the same channel, with the same polarization, is employed in the perpendicular paths but with different polarization in the parallel paths. The transmitters are alternately labeled Upper (U) and lower (L) duplex halves. Although the picture does not represent a physical ring configuration, the logical configuration and traffic flow are indeed ring in nature. If the ring consisted of an odd number of sites, there would be a conflict of duplex halves, and changing the frequency band would be a reliable alternative.

In the *star configuration,* as noted earlier, all transmitters on the common node should have the same duplex half (L). The configuration displays a difficult frequency planning scenario and is very sensitive to the geometry (mutual angles). If the node is a concentration point for high-capacity links, wide bandwidth is required, thus making the allocation of smaller channels in other portions of the network quite complicated. The link carrying the traffic out of the hub should use a frequency band other than the one employed inside the cluster.

Mesh configuration presents a complicated frequency planning scenario as a result of several conflicts of duplex halves. In addition, it normally requires more channels than do other configurations.

5.4 Microwave Design Tools

The International Telecommunications Union (ITU) publishes recommendations for the field of telecommunications. Recommendations for telecommunications are published in ITU-T, and recommendations that have been adapted for radio communication are published in ITU-R. The International Standards Organization (ISO) and American National Standards Institute (ANSI) are other organizations that promulgate standards, and they are referenced in this book where applicable.

For the proper planning of terrestrial line-of-sight systems, it is necessary to have appropriate and widely accepted propagation prediction models, methods, and data. Methods have been developed that allow the prediction of some of the most important propagation parameters affecting the planning of terrestrial line-of-sight systems. As far as possible, these methods have been tested against available measured data and have been shown to yield an accuracy that is both compatible with the natural variability of propagation phenomena and adequate for most present applications in system planning.

Most microwave network design software tools are developed by radio manufacturers and therefore are biased toward the manufacturers' own equipment. In other cases, the tool may be proprietary and not sold on the open market.[3] These tools are sometimes provided to engineering personnel who are working on the customer's site and performing network design. While some microwave equipment manufacturers insist on using their own software tools, some operators and consultants prefer to use commercially available tools. One such vendor-independent tool is PathLoss 4.0. This tool is probably one of the best (and most moderately priced) tools for the complex microwave design, including North American and ITU standards, different diversity schemes, diffraction and reflection (multipath) analysis, rain effects, interference analysis, and others.

Radio equipment parameters for equipment from any vendor, channel tables, antenna diagrams, and so on are defined and stored in the default parameters database for easy retrieval. Quality and availability calculations follow the ITU-R G.821 and G.826 recommendations. This tool is widely accepted by microwave system design engineers around the world.

5.5 Microwave Systems Engineering

5.5.1 System Documentation

In addition to the strictly microwave path engineering part of the project, there is a system engineering portion. System engineers usually provide technical direction and design to guarantee overall system integrity by verifying that all subsystems and contractor-furnished equipment are compatible, and that the desired performance is realized. Systems engineering will also provide transport traffic design and complete system integration as well as the network management system (NMS) and its integration into the other MW and fiber-optic systems.

In the next phase, application engineers review and translate the entire system configuration requirements into specific hardware implementations, including standardized interface levels, intra- and inter-rack cabling, and original equipment manufacturer (OEM) integration requirements, and they produce the following system documentation:

- Criteria, methods, standards, and procedures used for MW path engineering
- A system block diagram that details the main equipment provided by the contractor; all equipment grouped per site with interconnectivity between sites identified
- A block and level diagram that shows all the equipment provided by the contractor; may also show connections, interfaces, and signal levels to existing equipment at the same site
- A rack profile that shows the equipment mounting position on the rack
- A T1/E1 plan showing the system traffic routing; shows each T1/E1 connection from the originating site to the destination site
- A power consumption document that provides the value of the total power requirements of all equipment provided by the contractor according to the enclosed equipment spreadsheet, on a per-site-basis
- An equipment list (bill of materials) that includes all the equipment that needs to be provided
- Floor plans and equipment layout
- Wiring diagrams showing equipment and inter-rack wiring, cabling, waveguides runs, and so forth
- A channeling plan

- A synchronization plan
- Geographical system layout
- Path engineering results
- A tower profile showing all the radio equipment and transmission lines installed on the tower

5.5.2   Equipment Availability Calculations

A microwave link can become unavailable for a number of reasons, but this calculation includes only predictable equipment failures. There-fore, it excludes problems caused by misaligned or failed antenna feeder systems, extended loss of primary power, path propagation out-ages, human error, and other catastrophic events. Short-term (<10 sec) propagation outages are applied to the performance (not availability) objective and will not be used here. It is important to de-fine the terms and parameters used in equipment availability calcula-tions as follows:

$$A\ (\%) = 100\ (1 - U)$$

where A = availability (percentage of time, percent)
U = unavailability (percentage of time, percent)

For n pieces of equipment connected in series (tandem),

$$U = U_1 + U_2 + \ldots + U_n$$

For *two* pieces of equipment connected in parallel,

$$U = U_1 U_2$$

Unavailability of the microwave radio terminal can be expressed as:

$$U = 1 - \frac{MTBF}{MTBF + MTTR}$$

where $MTBF$ = mean time between failure ($MTBF = 10^9/FITS$)
$FITS$ = failures in time (10^9 hr)
$MTTR$ = mean time to repair

$$MTTR = RT + TT + (1 - P)TR$$

where RT = repair time on site
TT = travel time to the site
P = probability that a spare module is available
TR = time to obtain the spare module (assume 24 hr)

Example The typical protected MW terminal (1+1) has MTBF of 2,200,000 hr, it takes 0.5 hr to do the actual repair at the remote site, and the travel time is 3 hr. With good maintenance practice and spare parts available, we can assume P = 95 percent. Let us calculate unavailability of four MW hops connected in tandem.

$$MTTR = 0.5 + 3 + (1 - 0.95)24 = 4.5 \text{ hr}$$

Microwave hop (excluding all other equipment) has two terminals in series, so the unavailability is

$$U_{HOP} = 2\left[1 - \frac{2,200,000}{2,200,000 + 4.5}\right] = 0.409 \times 10^{-5}$$

For the four-hop system, total unavailability is:

$$U_{TOT} = 4 \times 0.000004091 = 0.000016364$$

Total availability is:

$$A_{TOT} = 99.99836 \text{ percent}$$

It is important to notice that this number includes only microwave terminals, and all the other equipment is excluded. Calculations that are more precise should include channel banks and multiplexers, power supplies, and other items.

5.6 Tips, Hints, and Suggestions

5.6.1 Basic Recommendations

In this chapter, we summarize some of the sound microwave network planning and design techniques that will help reduce the potential of having problems later during the system deployment. Most of the topics discussed here have been either mentioned and/or described in more detail elsewhere in this book, but it is a good idea to mention them again, in one place, in the form of a guidance checklist.

1. Use higher frequency bands for shorter hops and lower frequency bands for longer hops.

2. Avoid lower frequency bands in urban areas.

3. Use star and hub configurations for smaller networks and ring configuration for larger networks.

4. Use high-performance antennas in urban areas to minimize interference.

5. In areas with heavy precipitation, if possible, use frequency bands below 10 GHz.

6. Use protected systems (1+1) for all important and/or high-capacity links.

7. Leave enough spare capacity for future expansion of the system.

8. Space diversity is a very expensive way of improving the performance of the microwave link, and it should be used carefully and as a last resort.

9. The activities of microwave path planning and frequency planning preferably should be performed in parallel with line-of-sight activities and other network design activities for best efficiency. In addition, start the official process of frequency coordination and licensing process as soon as possible.

10. Use updated maps that are not more than a year old. Magnetic inclination as well as the terrain itself can change drastically in a very short time period. Make sure that everyone on the project is using the same maps, datums, and coordinate systems.

11. The datum or reference ellipsoid selection must be the same for the site data, image, and elevation files. The same map projection must be used for the image and elevation files.

12. Avoid using software programs with unknown algorithms. This could result in a microwave radio network with poor performance and quality.

13. Consider future network expansions in the frequency planning activity. It must be possible and easy to add new hops as the network expands.

14. Perform detailed path surveys on ALL microwave hops. Maps are used only for initial planning, as a first approximation.

15. Make sure that all of the obvious requirements are fulfilled (i.e., enough space for the transmission equipment, available AC and DC power, enough space for antenna installation and panning, access to the site, and so forth).

16. Below 10 GHz, multipath outage increases rapidly with path length. It also increases with frequency, climate factor, and average annual temperature. Multipath effects can be reduced with higher fade margins. If the path has excessive multipath outage, the performance may be improved by using one of the diversity methods.

17. It is very important to use an expert tower company to calculate the loading of the tower and the maximum allowed twist and sway of the structure. These decisions cannot be made on the basis of

qualitative perceptions or a "gut feeling." *Do not try to save money by using the "ballpark" method!*

18. Obtain microwave radio, transmission lines, antennas, and so forth only from a reputable and reliable supplier.

19. Waveguide installation is an extremely tricky operation—use only expert waveguide installers and riggers. People installing PCS and cellular antennas are not necessarily qualified for microwave antenna and waveguide installation.

20. Keep a good record of all the design documentation, survey reports, change orders, ATP results, and so on. Many of the long-term test results later will be useful as a benchmark for maintenance and troubleshooting.

21. In addition, do not try to overdimension the network, as this will make it unnecessarily costly. Every network, regardless of the type, will have brief outages from time to time, and microwave networks are no exception. *A network that does not fail is a fiction.*

5.6.2   Difficult Areas for Microwave Links

Some areas are more difficult for microwave links than others, and this is usually related to path or atmospheric conditions.[4] The following is a partial list of recommendations for the design and installation of microwave links in difficult areas:

1. In areas with lots of rain, use the lowest frequency band allowed for the project. Consult a local meteorological station for the "real" rain data and rely on the Crane or ITU rain maps only if no other information is available.

2. Be especially attentive during the design of microwave hops over or in the vicinity of the large water surfaces and flat land areas, as they can cause severe multipath fading. Reflections may be avoided by selecting sites that are shielded from the reflected rays.

3. Hot and humid coastal areas have a high ducting probability.

4. Desert areas may cause ground reflections, but sand does not have a high reflection coefficient. Most critical is the possibility of multipath fading and ducting caused by large temperature variations and/or temperature inversions.

5. Multipath typically occurs at sunrise, sunset, and during the night hours, when the air is calm, and stable refractive layers form in the atmosphere. Multipath is most common during the summer

months, when temperature and humidity differences are the most extreme. In North America, multipath outages are most severe near the Gulf of Mexico, the Great Lakes, and the Los Angeles basin, where humid air masses mix with dry continental air masses.

Multipath is much less common, for example, in the Rocky Mountain area, where air circulation in the mountains prevents the formation of stable atmosphere.

6. If upfading is a serious problem, smaller antennas, lower transmit power, or receiver attenuators can be used. These changes will improve upfade outage but can be used only on shorter paths, since this approach will reduce fading margins to combat multipath and rain. ATPC will reduce upfade outages without affecting the fade margin. If the angle of the path line to horizontal is more than 0.5°, upfading is not significant and can be ignored, since the microwave signal can penetrate the ducting layers that cause upfade.

5.7 References

1. Recommendation ITU-R P.452-7, *Prediction Procedure for the Evaluation of Microwave Interference between Stations of the Surface of the Earth at Frequencies above about 0.7 GHz,* 1995.
2. TIA, Telecommunications Systems Bulletin, TSB-10-F, *Interference Criteria for Microwave Systems,* 1994.
3. Lehpamer, H., *Transmission Systems Design Handbook for Wireless Networks,* Norwood, MA: Artech House, 2002.
4. Henne, I. and Thorvaldsen, P., *Planning of Line-of-Sight Radio Relay Systems,* Singapore: NERA Telecommunications, 2nd ed., 1999.

6

Microwave Deployment

6.1 Introduction

Microwave deployment (or implementation) is a multidisciplinary activity that involves a number of very specialized experts, regardless of whether it involves a new microwave system or an upgrade or expansion of the existing facilities. The related activities are as follows:

- Program/project management
- Site/path surveys (also part of design phase)
- Site civil work
- Site preparation, including grounding, lightning protection, and surge suppression
- Tower and building foundation construction
- Design, procurement, and erection of the antenna structures and equipment shelters
- Design, procurement, and installation of power systems (e.g., AC/DC/solar/diesel generators)
- Procure, install, integrate, test, and commission all the equipment required to complete the microwave transmission system
- Fulfillment of all regulatory requirements (e.g., local authorities, FCC, FAA)
- Completing as-built documentation
- Complete training (on-site and off-site), maintenance, technical support, and repair services
- Provision for the future upgrades and network expansion
- Testing (ATP) and commissioning

Not all the projects will include all of these activities. For example, the upgrade of an existing microwave system will not have tower erection included, but it may take in tower structural analysis and modifications or improvements as required.

6.2 Digital Microwave Radio

6.2.1 Microwave Radio Configurations

A *standard (all indoor) microwave radio configuration* consists of the entire microwave and digital modem part being placed indoors, the microwave antenna mounted outside on the tower, and a waveguide connecting the radio transceiver with the antenna. Today, the most commonly used waveguides for terrestrial microwave point-to-point systems are elliptical waveguides. This solution is acceptable for the lower frequencies, below 10 GHz, and high-capacity (backbone) microwave systems, but it quickly becomes unacceptable as frequency increases. This a result of the losses in transmission lines (coax or waveguide), which become unacceptably high at higher frequencies.

Whenever possible, *split configuration* microwave radio is replacing the standard configuration. To reduce losses between the transceiver and antenna, the outdoor unit (ODU) containing all of the RF modules can be mounted near the antenna. The ODU is connected to the indoor unit (IDU), which contains baseband circuitry, modulator, and demodulator, by means of a single coaxial IF cable. The distance between the indoor and outdoor equipment can usually be up to 300 m.

The equipment operates from a battery supply between −40.5 and −57 V, nominally −48 VDC. The primary DC power is supplied to the indoor unit through a main fuse and a filtering function, which includes input filter to attenuate the common mode noise. The power to the outdoor unit is supplied from the indoor unit via the IF coaxial cable.

6.2.2 Basic Microwave Radio Parameters

6.2.2.1 Transmit output power. Transmit output power is RF power, usually expressed in decibels referenced to a milliwatt (dBm). It is always necessary to define the interface related to the transmit output power value (e.g., antenna port or radio port). Without this information, it is impossible to calculate the real gain/loss of the radio equipment. If the radio includes a booster or a high-power amplifier or has an adaptive transmit power control (ATPC) implemented, this factor must be considered in the transmit output value. In addition, if trans-

mit output power can be reduced (e.g., –3 or –6 dB below its nominal value), one must also be aware of it and reduce interference levels to acceptable limits.

6.2.2.2 Transmit frequencies. Transmit frequencies, expressed in giga- hertz, include the remote site transmit frequency and the local site transmit frequency. In most microwave radio systems, nonintrusive monitor points are available for the measurement of RF output power and transmit and receive local oscillator (LO) frequencies. Therefore, the measurement of transmitter power, along with LO frequencies and power, are recorded on a routine maintenance basis.

6.2.2.3 Equipment configuration. The equipment configuration can be either nonprotected or protected. Several types of protection schemes and diversity arrangements are described in other chapters of this book.

6.2.2.4 Link polarization. The microwave radio link polarization must be of linear type, either horizontal (H-pol) or vertical (V-pol).

6.2.2.5 Antenna data. This includes the model and manufacturer, di- ameter (expressed in meters or feet), and mid-band gain expressed in dBi or dBd. Terrestrial microwave antenna systems use parabolic an- tennas in most cases. However, sometimes other types of antennas may be utilized, such as horns and, in the beginning microwave fre- quencies, Yagis and helicals.

6.2.2.6 Bit rate. Bit rate is usually expressed megabits per second (Mbps). Gross bit rate (GBR) is the final result of the bit rate after in- serting the forward error correction (FEC), voice frequency channel, data channel, and control/monitoring channel via the radio equip- ment.

6.2.2.7 Receive noise. Receive noise, expressed in decibels (dB), is the noise measured at the receiver input port.

6.2.2.8 Transmit spectrum mask (spectrum occupancy). Spectrum anal- ysis is one of the most important measurements in digital radio test- ing. The spectral occupancy test is a measure of how well unwanted sideband and spurious signals have been suppressed by successive fil- ters in the transmitter. Digital microwave radios operate with well de- fined and controlled spectral occupancy; therefore, it is routine practice to measure the occupancy of the radio against predefined limit or masks.

All filters and devices (e.g., circulators, isolators, transitions, elbows) must be right in the beginning of the waveguide (elliptical, circular, or rectangular) or coaxial cable in lower frequencies. They must be provided in the form of a decibel-versus-frequency-offset curve. The offset represents the positive or negative frequency offset from the carrier frequency ($f_0 + \Delta f$ and $f_0 - \Delta f$) and is generally expressed in megahertz. In terms of unwanted emissions (spurious and out of band), the equipment must meet the appropriate specifications (e.g., in the U.S., FCC Part101, Section 101.111, Emission Limitations).

6.2.2.9 Receive frequency response. The receive chain (RF + IF + BB) filtering is composed of three curves: the receive RF filter frequency response curve, the receive IF filter frequency response curve, and the receive BB filter frequency response curve. All three curves show the filter response (in decibels) versus the frequency offset (in megahertz).

6.2.2.10 Attenuator and additional losses (expressed in decibels). When a fixed in-line attenuator (pad) is used, it is necessary to recognize whether the attenuator belongs to the transmit path only, to the receive path only, or to both paths. Additional losses are divided into two parts, and they both include the portion between the antenna interface and the radio interface. The antenna interface is located between the antenna and the transmitting line (waveguide or coaxial cable). The radio interface is located between the radio equipment and the transmitting line. The antenna interface consists of the transmit path losses, and the radio interface consists of the receive path losses. Depending on the situation, there could be some differences between these two losses.

Overall additional losses are equal to the summation of losses in the transmission line, flex-twists (or jumpers, for coaxial), hybrid devices, switches, circulators, isolators, straight sections, flange adapters, waveguide-to-coaxial adapters, taper transitions, 90° E-plane and H-plane elbows, power dividers, and the radome.

6.2.2.11 Receiver sensitivity threshold. The receiver sensitivity or threshold (Rx) defines the minimum signal strength required for a radio to successfully receive a signal. Receiver sensitivity is a function of the receiver's noise factor, the noise bandwidth, and the modulation method. A radio cannot receive or interpret a signal that is weaker than the receiver sensitivity threshold. The receiver threshold is the receive power level normally at the antenna interface (equal to antenna port) for a given BER (bit error rate). Receiver thresholds expressed in dBm for a 10^{-3} BER in PDH radios, and 2×10^{-5} BER in SDH radios, are usually provided.

6.2.2.12 Receive signal level. The receive signal level (RSL) is the expected strength of a signal when it reaches the receiving radio. The following formula defines the RSL:

$$RSL = P_o - L_{ctx} + G_{atx} - L_{crx} + G_{atx} - FSL$$

where P_o = output power of the transmitter (dBm)
L_{ctx} = cable loss between the transmitter and its antenna (dB)
G_{atx} = gain of the transmitter's antenna (dBi)
L_{crx} = loss (cable, connectors, branching unit) between the receiver and its antenna (dB)
G_{atx} = gain of the receiver's antenna (dBi)
FSL = free-space loss (dB)

6.2.2.13 Link feasibility formula. To determine if a link is feasible, compare the calculated receive signal level with the receiver sensitivity threshold. The link is theoretically feasible if

$$RSL \geq Rx$$

6.2.2.14 C/N vs. BER. The noise and interference test set is used with a bit-error rate test (BERT) to make one of the most common and important measurement on a digital microwave radios, i.e., the C/N curve. This measurement is made at virtually every stage of a radio's development, production, and use. Later measurements can be made to identify faults or gradual degradation in performance.

6.2.2.15 Fade margin and link availability. Fade margin is the difference between the unfaded RSL and the receiver sensitivity threshold (Rx). Each link must have sufficient fade margin to protect against path fading that weakens the radio signals. Fade margin is the link's insurance against unexpected system outages and is directly related to link availability, which is the percentage of time that the link is functional. The percentage of time that the link is available increases as the fade margin increases, while the link with little or no fade margin may experience periodic outages.

6.2.3 Radio Performance Improvement

Microwave radios take advantage of a number of powerful antifading measures. These include space diversity reception, ATPC, adaptive equalizers, multilevel coded modulation (MLCM)-type forward error correction (FEC), cross-polarization interference cancellers (XPICs), high-speed errorless switching with early warning, and others.

6.2.3.1 Microwave link protection. The terms *protection* and *diversity* are often used interchangeably when applied to microwave links. This is incorrect, since protection commonly improves long-term traffic interruptions (10 CSES or more), whereas diversity arrangements greatly reduce the number and duration of short-term outages (less than 10 CSES). (CSES is the abbreviation for *consecutive severely errored seconds*).

The monitored hot standby (MHSB) has two transmitters and two receivers that are always on-line ("hot"). A switch keeps one radio transmitting or receiving until a failure occurs and, at that moment, the signal is switched to the standby radio. The monitored hot standby (MHSB) configuration (1+1) protects against equipment failures only, not path propagation problems. The hot stand-by protection system uses a common frequency channel with only one of the two transmitters at any end of the microwave link active at any particular time. The transmitter in standby mode will be fully operational except that the transmitter will be muted. The hot standby (protected) system configuration provides hitless receiver changeover on each side of the radio relay link in case of receiver equipment failure or sudden propagation path fading on one of the four microwave paths. If a transmitter fails, there will be a short break in transmission until the standby transmitter is activated. In contrast to receiver changeover, transmitter changeover therefore will not be hitless.

Nonprotected systems (1+0) consist of one indoor unit and one outdoor unit interconnected with a single coaxial cable. In the case of failure of any of the electrical or mechanical components, the entire microwave hop will fail.

In hot standby (protected) configuration (1+1), the IDU, the transceiver unit, and the coaxial cable between IDU and ODU are duplicated. The two transceivers share the same branching unit. A switch at radio frequency level, included in the branching unit, allows for switching between the two transmitters. A splitter and switching unit placed between the two IDUs are added. For each terminal, one antenna is used. In addition, there is a version of the protected system with two ODUs and only one antenna.

Split-configuration radios were designed with the ease of installation in mind and to eliminate the waveguide. The equipment can be installed within a few hours. These radios can also be fully indoor mounted if necessary. One coaxial cable between the IDU and ODU is used for "1+0" systems, and two cables for hot standby systems (1+1).

6.2.3.2 Adaptive equalizers. Critical to performance under multipath conditions, microwave radio demodulators have an adaptive equalizers. These powerful equalizers greatly reduce many of the degradations caused by the radio and the microwave path and therefore allow longer microwave hops. Adaptive equalizers are either spectrum-driven (adaptive slope amplitude equalizers, or ASAEs) operating in the frequency domain or decision-feedback devices (adaptive time domain equalizers, or ATDEs) operating in time domain. Either or both of these equalizers is incorporated into higher-capacity digital radios as required to meet the long-term performance (outage and quality) objectives over a wide range of atmospheric fade characteristics, path geometries, equipment protection schemes, temperatures, and other characteristics.

6.2.3.3 Forward error correction. Forward error correction (FEC) is an error correction scheme that adds redundant bits to the payload input to the digital transmitter, thereby increasing transmitted symbol rate and RF bandwidth so as to correct random errors at the receiving terminal. The wider spectrum bandwidth is, on the other hand, more vulnerable to dispersive fade outages. FEC corrects low BERs or so-called "dribbling" or random errors, and it is not effective for BER $< 10^{-4}$.

6.2.3.4 Cross-polarization interference canceller (XPIC). With dual-polarity transmission, two signals are transmitted—one with horizontal polarity and one with vertical polarity. The main challenge with this kind of configuration is cross-polarization interference, whereby energy from one polarization is received in the other. Cross-polarization interference is the result of equipment imperfections, the link environment (atmosphere, terrain), and the impact of rain. This could lower the cross coupling (the cross-polarization discrimination, XPD) between the co-channel dual-polarized signals on the main and diversity channels simultaneously to an unacceptable value, thus degrading the link's error performance.

The cross-polarization interference canceller (XPIC) feature is a technology that allows transmission on both the horizontal and vertical polarity of any given frequency pair. This way, the maximum link capacity will effectively double for a given frequency band. Such cancellation is essential to achieve performance that is not limited by cross-polarization interference but rather by the co-polarization attenuation, as with single polarity radios. Improvement achieved by using XPIC is typically around 20 dB.

XPIC design considerations are as follows:

- Ultra-high performance, dual polarized antennas have to be used. Dual polarized antennas have two feed horns that transmit the microwave energy in the vertical and horizontal planes.

- The antenna discrimination between the two poles is typically 30 to 40 dB, which reduces adjacent channel interference.

- It is preferable to have ATPC operational (both directions) on the XPIC path.

- For circulator coupled XPIC equipment, an additional loss for the transmitters and for the receivers must be taken into account in the path calculations. These losses can be sometimes quite high (6 to 8 dB).

- XPIC should be considered in the initial system design; retrofitting the equipment in the field is almost impossible after it is installed.

6.2.3.5 Adaptive transmit power control (ATPC). The radio output power can be controlled in fixed or adaptive mode. In *fixed mode,* the output power P_{out} ranges from a minimum level $P_{fix\ min}$ to a maximum level P_{max}. A value P_{set} is manually set in 1-dB increments, locally or remotely from the management system.

In *adaptive mode* the automatic transmit power control (ATPC) function is used to automatically control the output power, P_{out}. The output power is continuously adjusted so as to maintain a minimum input level set from the far-end terminal. Under normal path conditions, the ATPC maintains the output power at a reduced level, resulting in a lower interference level in the radio network.

ATPC is a feedback control system that temporarily increases transmitter output power during periods of fading, thus eliminating (or at least reducing) the adverse effects of fade events on digital point-to-point microwave fixed services. ATPC offers immediate and long-term advantages to the link operator, including reduced average power consumption, extended equipment MTBF, and lower long-term RF interference levels. Propagation statistics indicate that fade events on physically different propagation paths are noncorrelated; thus, the probability of simultaneous sensitivity to interference for two separate systems is small—at least for situations in which multipath fading is the dominant limiting factor. As long as link paths are properly designed with adequate path clearance and are not significantly affected by rain fade events, the ATPC maximum transmit power boost is required only for appropriately short periods of time (<2 sec). Transmit power in excess of coordinated power is maximum 10 dB and is allowed for not more than 0.01 percent of the time (3,250 sec/year).

There are two main advantages to using ATPC: the transmit power less than the maximum power may be used for the calculation of interference into other systems, and calculations of interference into the receiver of a system using ATPC may assume that the wanted signal transmitter is operating at maximum transmit power.

6.2.3.6   Channel width, spectral efficiency, and modulation schemes.   A sinusoidal electromagnetic wave has three properties: amplitude, frequency, and phase. Any one of these parameters can be modulated to convey information. The modulation may be either analog or digital. In analog signals, the range of values of a modulated parameter is continuous. In terrestrial radio systems, for example, AM and FM channels represent amplitude and frequency modulation, respectively. In digital signals, the modulated parameter takes on a finite number of discrete values to represent digital symbols. The advantage of digital transmission is that signals can be regenerated without any loss or distortion to the baseband information.

By far the most common form of modulation in digital communication is M-ary phase shift keying (PSK). With this method, a digital symbol is represented by one of M phase states of a sinusoidal carrier. For binary phase shift keying (BPSK), there are two phase states, 0° and 180°, that represent a binary one or zero. With quaternary phase shift keying (QPSK), there are four phase states representing the symbols 11, 10, 01, and 00. Each symbol contains two bits. A QPSK modulator may be regarded as equivalent to two BPSK modulators out of phase by 90°.

Some other modulation schemes commonly used in data systems are FSK (including BFSK and QFSK) and QAM (of various levels).

A suitable modulation method is selected by taking into account the system requirements. For instance, if spectrum efficiency is not a major issue and/or high interference tolerance is important, a simple modulation method should be used. *Spectral efficiency* refers to spectrum utilization as measured in bits/sec/Hz, and some countries and their regulatory organizations do not allow microwave radios to be installed if they do not have a very high spectral efficiency. The features of simple modulation methods are as follows:

- Easy implementation in all frequency bands
- Robustness against propagation effects
- High tolerance against all kinds of interferences
- High system-gain characteristics

On the other hand, a multistate modulation method improves spectral efficiency on a route. Sophisticated modulation methods (high levels of QAM and TCM) are required for high-capacity links so that more information can be packed together. Typical applications for these multistate modulation methods are high-capacity trunk, junction, and access networks. The method of modulation and the capacity will directly affect the required bandwidth and the interference tolerance of a radio link.

The unit of information is the *symbol*, and the different schemes use different numbers of bits to define each symbol. For a given symbol rate, the greater the number of bits per symbol, the higher the data rate.

For example, 8PSK has eight states in steps of 45°. Shifting the carrier phase by 45° requires 1 Hz of the carrier frequency for 3 bits of the base band (23=8), and the spectral efficiency is 3 bps/Hz. A 2-Mbps baseband modulated with 8PSK requires an RF carrier with a bandwidth of 0.67 MHz. A 140-Mbps baseband therefore requires an RF carrier that has a bandwidth of 47 MHz.

On the other hand, 64QAM has 64 states, and phase and amplitude are shifted. One shift requires 1 Hz of the carrier frequency for 6 bits of the base band (26=64), and the spectral efficiency is 6 bps/Hz. A 2-Mbps baseband modulated with 64QAM requires an RF carrier with a bandwidth of 0.33 MHz. A 140-Mbps baseband requires an RF carrier that has a bandwidth of 23 MHz.

As shown in Table 6.1, the STM-1 radio link needs the radio channel 112 MHz with 4QAM, 56 MHz with 16QAM, and 28 MHz with

TABLE 6.1   Channel Requirements for Different Modulation Methods

Modulation	Bandwidth (16 × 2 Mbps)	Bandwidth (155 Mbps)	S/N (10^{-6})	System gain (relative to 4QAM)
4QAM	28 MHz	112 MHz	13.5 dB	0 dB
16QAM	14 MHz	56 MHz	20.5 dB	7 dB
32TCM-2D	14 MHz	56 MHz	17.6 dB	4 dB
64QAM	14 MHz	56 MHz	26.5 dB	13 dB
128QAM	7 MHz	28 MHz	29.5 dB	16 dB
256QAM	7 MHz	28 MHz	32.6 dB	19 dB

128QAM. The S/N requirements for receiver threshold using different modulations are also shown in the table. For example, 128QAM needs about 16 dB higher S/N than 4QAM. To upgrade the existing radio link from 4QAM 16×2 Mbps to 155 Mbps without changing the RF-channel width, 128QAM is needed. A transmit power increase is normally not possible ($P_{tx\,max}$ < 30 dBm) beyond more than few decibels (0 to 3 dB).

The required increase in power to attenuate noise and interference for 16QAM, 64QAM, and 128QAM compared to 4QAM is 7, 13, and 19 dB, respectively. The result is that these modulation methods with high number of states are very inefficient in dense city networks containing many randomly oriented links (e.g., a mesh network).

If the size of antennas in both ends is doubled, roughly 6+6 dB can be gained. Due to practical reasons, antenna gains exceeding about 44 dB cannot be used. With these changes, 12 to 15 dB is possible and, in case of short hops, the missing part may be "covered" by the excess system gain margin.

If the RF-channel can be changed from 28 to 56 MHz, this kind of capacity upgrade would be less critical, because 32TCM can be used. The system gain increase demand would then be only 4 dB and could be partly covered by transmit power increase (e.g., 3dB) or with a 50 percent bigger antenna at one end. If antenna changes are planned, the rigidity and available space of the supporting structures must be checked. If antenna changes are not possible, the hop lengths normally must be reduced.

The general rule is that, to obtain higher data rates, higher-order modulation schemes must be employed. Sophisticated modulation methods for high SDH capacity links can improve the spectral efficiency, but the price paid for this increased throughput is an increase in operating threshold level (and more susceptibility to interference). In real terms, the trade-off is data rate versus energy per bit, and thus range. Since an increase in bandwidth normally introduces an increase in error rate, error-correcting schemes (overhead) are often included to produce a net gain in data throughput.

Detailed information on the subject of spectrum efficiency is provided in Recommendation ITU-R SM.1046.

6.2.3.7 Coded modulation. This is a technique that combines coding and modulation that would have been done independently in the conventional method. Redundant bits are inserted in multistate numbers of transmitted signal constellations. This process is known as *coded modulation.* Representative examples of coded modulation are block coded modulation (BCM), trellis coded modulation (TCM), and multi-level coded modulation (MLC or MLCM). In BCM, levels are coded by block codes, whereas TCM uses only convolutional codes. On the other hand, different codes can be used for each coded level in MLCM, so MLCM can be seen as a general concept that includes BCM and, to some extent, TCM. These schemes require added receiver complexity in the form of a maximum likelihood decoder with soft decision.

A technique similar to TCM is the *partial response,* sometimes called a *duo-binary* or *correlative* signaling system. A controlled amount of intersymbol interference, or redundancy, is introduced into the channel. Hence, the signal constellation is expanded without increasing the transmitted data bandwidth. There are various methods utilizing this redundancy to detect and then correct errors to improve performance. This process is called *ambiguity zone detection or AZD*.

6.2.3.8 Receiver data switching. *Hitless receiver data switching* is the most commonly used diversity switching method in microwave radios today. This type of switching between receivers may contribute to additional errored bits, but no hits (service disruptions) occur. Although sufficient in most cases, in a severe fade environment, as well as on long multihop system, it can create enough errors to cause problems.

Errorless receiver data switching between receivers is achieved with some type of anticipatory sensing, usually at a BER of 10^{-8} or before errors occurs. This is the preferred configuration for high-capacity links carrying critical information over a long, multihop microwave system and in a high multipath fade environment with common diversity switching activity.

The caution is required in systems in which FEC is utilized, since it can cause problems or completely disable the hitless/errorless switch feature.

6.2.4 T/I Curves

The T/I curves of the radio are used for interference analysis and frequency coordination, and they are based on lab measurements. The T/I is considered the minimum interference level that will have any significant effect on the radio. It is determined by first fading the radio to the 10^{-6} BER static threshold and then adding an interfering signal until the static threshold is degraded by 1.0 dB. The difference between the interfering signal level and the threshold is defined as T/I. By changing the frequency separation between the interfering signal and the receive frequency, a curve of T/I values is produced. Typically, two different T/I curves are generated for each radio, one using a modulated digital signal as the interferer and the other using a CW tone as the interferer. The modulated curve is used to analyze digital interference cases, and the CW curve is used for FM cases.

The digital T/I curve assumes like modulation, and it may not be accurate if the interfering transmitter is not the same radio type as the victim receiver.

6.2.5 Service Telephone Network

A service telephone network can be created using the *engineering order wire* facilities provided on the SDH equipment. An operator who is seeking to maintain a quality network will require access to all network sites from a central point or from more than one site. There are, in general, two ways to achieve communication to sites. The first is to use the PSTN network and provide a telephone line to each site. The second is to set up a *service telephone network* for each site.

A service telephone network is a system that gives the operator access to sites where there is transmission equipment. Within the design of the SDH equipment is a facility called *engineer order wire (EOW)*, allowing a telephone to be connected to the equipment and enabling voice communication via an AUX card.

To achieve this, the voice channel is carried by the "E1" or "E2" bytes in the SDH section overheads. These bytes are dedicated to voice (EOW) by the ITU-T. The EOW is always connected in a party-line network, which means that all NEs configured to have access to E1 or E2 can have simultaneous access to the service telephone. A protocol is used to provide selective or omnibus calls. The EOW interface is a standard analog 2w interface, DC current feeding, hang off status detection, and DTMF signaling.

6.2.6 Duplexers

The duplexer is the main component that can affect the manufacturing time of the microwave radio and therefore the lead times for the microwave network deployment. Any two-way wireless communication requires both a transmitter and a receiver and, in a full duplex operation, they both operate at the same time. The duplexer is an antenna-coupling device that allows a transmitter and a receiver to be connected simultaneously to the same antenna (see Fig. 6.1).

Even if each had its own antenna, full-duplex operation can present a problem, because the power output of the transmitter is greater than the power level of signals the receiver is trying to receive. There-

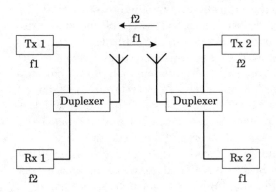

Figure 6.1 Full duplex traffic with simultaneous transmission.

fore, when these two devices are operating at the same time in close proximity, some of the energy from the transmitter will find its way into the receiver where it will surely be more powerful in comparison to the signals the receiver wants to receive.

When the transmitter and receiver are connected to the same antenna, the problem becomes even more acute. For full duplex to work at all, it is necessary to transmit and receive on different frequencies. This system is called *frequency-division duplex*. The idea is that the receiver will not be able to "hear" the transmitted signal, because the receiver is selective; it will receive only a frequency (or a small range of frequencies) to which it is tuned and will not receive the transmitted signal if the frequency is outside of the receiver's tuning range (called the *receive passband*).

Although this fundamental idea sounds simple, there can still be a problem. The receiver obtains its selectivity characteristic by using filters, which pass certain frequencies while rejecting others.

The duplexer can be thought of as just a pair of bandpass filters incorporated together in one box. It has three connection ports: the transmit (Tx) port, the receive (Rx) port, and the antenna port. The Tx and Rx ports are usually interchangeable and, in most implementations, the duplexer is a passive device—meaning it neither requires nor consumes any power. Consequently, the duplexer is not user configurable through software control or any other means. Mechanical adjustments that are made at the time of manufacture should never need readjustment or calibration.

The duplexer will have two nonoverlapping passband frequency ranges, and thus one will naturally be higher than the other. It is possible to set up a system to transmit through the higher frequency passband filter and receive through the lower frequency one, or vice versa. These two scenarios are usually described as *transmit high* or *transmit low*. The only real requirement is to make sure that the transmit frequency falls within the passband range of one of the duplexer's filters, and the receive frequency falls within the other. This requires the operator to know the passband frequency ranges of the duplexer and the Tx and Rx operating frequencies when ordering microwave equipment. In practice, one first will determine, to at least a rough degree, what the transmit and receive frequencies will be. Then, a duplexer is chosen with appropriate Tx and Rx passband ranges to accommodate the desired operation frequencies

After the system is installed or ordered, if it is desired to alter either the Tx or Rx frequencies (or both), this can be done as long as any new chosen frequencies fall within the duplexer's passbands. Otherwise, it will be necessary to obtain a different duplexer (for each end of the link).

6.2.7   Environmental Requirements

Every microwave project is different, and the environmental requirements for the transmission equipment used can vary from country to country and from project to project. Therefore, there is a chance that the equipment will have to be tested and certified to make sure it complies with local standards and regulations as well to make sure it will survive harsh environmental conditions.

In case the MW radio is designed in split (indoor-outdoor) configuration, the MW radio's ODU will be used in an outdoor environment. External temperatures could vary between −40 and +50°C. In that case, the enclosure should be internally insulated with a suitable insulation material to minimize interior temperature variations caused by surface heat absorption, and it should be equipped with a suitable temperature and environmental (cooling and heating) control.

The ODU enclosure itself should conform to North American NEMA 4 or equivalent European IEC 529 standards (IP55, Protection from Water, Snow, Dust Entry). If possible, testing should be done by a third party. The equipment supplier should supply test results from an independent test lab, including its detailed environmental testing specification. The ODU will be used for housing electronic equipment and should be shielded from dust, moisture, and any external contaminants that could enter it. The ODU should also be corrosion resistant.

For indoor applications where HVAC is provided, the transmission/microwave equipment shall indefinitely satisfy all manufacturers' published specifications within a temperature range of 0 to +45°C (ideally −10 to +45°C). For both outdoor and indoor applications, the equipment shall indefinitely satisfy all manufacturers' published specifications within a humidity range of 0 to 90 percent noncondensing relative humidity. All microwave equipment must satisfy manufacturers published specifications while operating in an altitude pressure density range of −100 to +4,000 m (above mean sea level pressure). All microwave radio specs have to be guaranteed over the entire temperature range.

6.2.8   Network Management System

The ability to remotely manage equipment from a central management site is an essential requirement of operating any communication network, and microwave networks are no exception. The ability to fully report on and diagnose radio equipment and its performance is very important, since deploying technicians to visit remote tower sites is time consuming and costly.

Local management (e.g., configuration and setup, software download, and so forth) at site is performed from the *local craft terminal*

(LCT), which allows access to all operation and maintenance facilities on the terminal via a web server in the terminal.

Remote management can be performed over a DCN from the appropriate NMS. Remote supervision allows monitoring of alarms and performance as well as some configurations. Remote supervision of the SDH microwave network is realized with a connection to one of the terminals in the network. The terminal unit is usually equipped with a 10BASE-T port for this purpose. Each terminal also holds a router that terminates and routes IP messages, which means that the DCN can be extended throughout the transmission network. The router ensures that O&M data on the 10BASE-T port are incorporated in the main traffic over the hop, using either the EOC or DCC channel.

For exchange of O&M data between two terminals on the same site, the units can be interconnected by the 10BASE-T ports through an external hub (alternative 1 in Fig. 6.2). Another possibility is to let O&M data pass unaffected between the terminals in the EOC/DCC channel (alternative 2).

Fault management deals with detection, isolation, and correction of malfunctions. It can be used with performance management as well, to compensate for environmental changes. Also included are maintenance and examination of error logs and action on alarms.

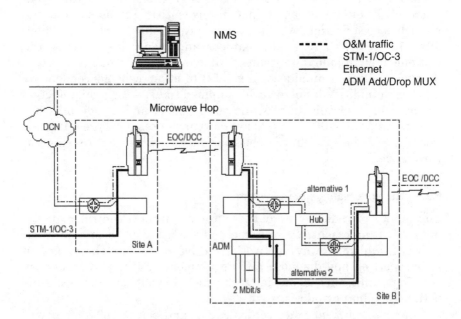

Figure 6.2 Routing O&M traffic in an SDH microwave network.

Performance management can be described as a set of functions to evaluate and report the behavior of the equipment and to analyze its effectiveness. It also includes subfunctions to gather statistical information. Performance management is based on the ITU-T G.826 and M.2120 recommendations.

The built-in *security functions* protect the terminal and its services from disturbances caused by illegal activities of nonauthorized personnel. The access control is carried out when the user accesses the terminal from the LCT or when the Java-based user interface is started from the NMS. The user has to log on to the management system with a user name and a password to be able to access the management facilities. Depending on the selected mode, the user will be able to obtain read-only access or read/write access to the system.

6.2.9 Microwave Compatibility and Safety

In addition to the obvious concerns, such as safety when climbing structures or while working with dangerous AC line voltage, there is also the issue of exposure to RF radiation. Much is still unknown, so there is a great deal of debate concerning the safe limits of human exposure to RF radiation.

The American National Standards Institute (ANSI) sets safety standards for human exposure to radio frequency (RF) electromagnetic energy in the United States. Exposure standards for RF energy are threshold standards. Unlike ionizing radiation, which many people believe to act cumulatively, even at low exposure levels, RF exposure at low levels is not considered to be a cumulative hazard. Threshold standards define the level of RF energy above which there may be health hazards and below which there have been no reported harmful effects. ANSI conservatively set its maximum permissible exposure levels for RF energy at one-tenth (or less) of the threshold for human health effects. The maximum permissible exposure levels for protection against RF energy recommended by ANSI are comparable to those set in other countries. Government agencies recognize and generally accept the ANSI RF safety standard (ANSI/IEEE C95.1). For more information see Ref. 1.

Ionization is a process by which electrons are stripped from atoms and molecules. This process can produce molecular changes that can lead to damage in biological tissue, including effects on the genetic material, DNA. This process requires interaction with photons containing high energy levels, such as those of X-rays and gamma rays. A single quantum event (absorption of an X-ray or gamma-ray photon) can cause ionization and subsequent biological damage due to the high energy content of the photon, which would be in excess of 10 eV and considered to be the minimum photon energy capable of causing ionization. Therefore, X-rays and gamma rays are examples of ionizing radiation. Ionizing radiation is also associated with the generation of nuclear energy, where it is often simply referred to as *radiation*. The photon energies of RF electromagnetic waves are not great enough to cause the ionization of atoms and molecules, so the RF energy is characterized as *nonionizing radiation,* along with visible light, infrared radiation, and other forms of electromagnetic radiation with relatively low frequencies. It is important that the terms *ionizing* and *nonionizing* not be confused when discussing biological effects of electromagnetic radiation or energy, since the mechanisms of interaction with the human body are quite different.

The best general rule is to avoid unnecessary exposure to radiated RF energy. This means not standing in front of, or in close proximity to, any antenna that is radiating a transmitted signal. Of course, antennas that are used only for receiving do not pose any danger or problem. For dish-type antennas, it is safe to be near the operating transmit antenna if you are at its back or side, as these antennas are directional, and potentially hazardous emission levels will be present only in front of the antenna.

Always assume that any antenna is transmitting RF energy, especially since most antennas are used in duplex systems. Be particularly wary of smaller dishes [1 ft (0.3 m) or less], as these often radiate RF energy in the higher frequency ranges. The general rule is that the higher the frequency, the more potentially hazardous the radiation. It is known that looking into the open (unterminated) end of waveguide that is carrying RF energy at 10 GHz or more will cause retinal damage if the exposure lasts as little as tens of seconds at a transmit power level of only a few watts. In any case, be careful to ensure that the transmitter is not operating before removing or replacing any antenna connections.

On a rooftop and near an installation of microwave antennas, it is important to avoid walking, and especially standing, in front of any of the equipment. If it is necessary to traverse a path in front of any such antennas, there is typically a very low safety concern if you move quickly across an antenna's path axis.

More information can be found in the literature.[2-4]

6.2.10 Microwave Radio Installation

Installation of the split-terminal configuration, assuming that all the infrastructure is available (tower, shelter, racks, AC/DC power, pipe-mounts, and so on), should not take more than one day per hop and two people to do the work. Indoor installation consists of one small module for nonprotected systems and three modules for protected systems (two modems plus the switching module). The usual telecommunications rack consists of the fuse or breaker panel on the top, space for the small rectifier and battery backup at the bottom of the rack, and microwave radio and the DSX panel mounted in the middle of the rack.

Standard microwave radio installation can take few days, since they are usually bigger in size and have rigid waveguide components, flexible waveguides, flanges, dehydrators, and so forth. They usually need additional bracing and specialized tools to mount and make them operational.

The installation team has to have all the necessary tools, installation equipment, and test equipment to complete the project. The basic list of test equipment is as follows:

- A bit error test set, covering the data rates, interfaces, and protocols of the equipment to be installed
- Power meter(s) and associated power sensors covering the frequency bands of the microwave radio equipment
- Frequency counter(s) covering the frequency bands of the microwave radio equipment
- Variable attenuators covering the frequency bands and matching the flange/connector type of the radio equipment
- Multimeter(s)

6.3 Digital Multiplexers

Today, digital multiplexers are part of every microwave system, whether they are part of the microwave radio equipment or added as an external piece of equipment. *Multiplexing* is a process in which multiple data channels are combined into a single data or physical channel at the source; conversely, demultiplexing is the process of separating multiplexed data channels at the destination. An example of multiplexing is when data from multiple devices is combined into a single physical channel. Some methods used for multiplexing data are *time-division multiplexing (TDM), asynchronous time-division multiplexing (ATDM), frequency-division multiplexing (FDM),* and *statistical multiplexing.*

One method (FDM) involves splitting the frequency band transmitted by the channel into narrower bands. Each of these narrow bands is used to create a distinct channel. In FDM, information from each data channel is allocated bandwidth based on the signal frequency of the traffic, and multiple channels are combined onto a single aggregate signal for transmission. The channels are separated by their frequency. FDM was the first multiplexing scheme to be widely used, and such systems are still in use. However, TDM is the preferred approach today. In TDM, information from each data channel is allocated bandwidth based on preassigned time slots, regardless of whether there is data to transmit. In ATDM, information from data channels is allocated bandwidth as needed, using dynamically assigned time slots. In statistical multiplexing, bandwidth is dynamically allocated to any data channels that have information to transmit.

M13 is a most commonly used multiplexer and usually takes the form of a modular, compact unit for multiplexing up to 28 DS1s into a DS3. Compatible with North American standard interfaces, these multiplexers are capable of multiplexing up to seven low-speed DS1 signal groups (each with four DS1s) into one DS3 (44.736 Mbps) signal, with each low-speed signal group consisting of four DS1s. Standard features include continuous performance monitoring and extensive local and remote diagnostics, which provide off-premises restoration, thereby avoiding the cost of on-premises maintenance. Remote monitoring and control are the key features, e.g., remote provisioning, remote inventory, performance monitoring, and remote testing.

FMT150 is a fiber-optic multiplexer that multiplexes and transports one, two, or three DS3 signals; up to 84 DS1s; and a maximum of 2,016 voice circuits. Designed to interface with DS1 or DS3 input signals, the typical FMT150 uses 150 Mbps fiber transport plug-ins with up to three DS3 inputs. FMT150 usually features a comprehensive maintenance system and provides complete, instantaneous monitoring and troubleshooting capabilities, including fault analysis of the remote site. It can also monitor alarms and controls at local sites and provide so-called *housekeeping alarms*.

The purpose of *sub-rate multiplexing* is to fit more sub-DS0 data circuits into one DS0 (64-kb) channel. The SRDM feeds synchronous or asynchronous data circuits into a single 64-kbps DS0 signal. For example, the SRDM can multiplex 20 sub-DS0 signals into a DS0 channel at 2.4 kbps, or 10 at 4.8 kbps, or 5 at 9.6 kbps. Sub-rate multiplexers were very popular during the early TDMA and GSM wireless networks build-out, since they allowed circuit grooming and efficient use of T1/E1 facilities.

Drop and insert multiplexers allow a transiting DS3 signal to drop and insert a small number of DS1s without being fully demultiplexed and remultiplexed as would occur with back-to-back terminals.

In the case of a *regenerative repeater* site where mux was not provided, most microwave radios themselves have add/drop capability of one or more DS1 circuits. In this case, additional multiplexer may not be required.

6.4 Cabling and Signal Termination

There is very small chance that a microwave engineer working on the design and deployment of a microwave network will get involved in the details of the metallic or fiber-optic cable installation. Microwave engineer may be involved in routing of the coax cable connecting outdoor and indoor units of the split-configuration microwave radio. In addition, interfacing the existing metallic or fiber-optic network at the switch office location and hub sites is very likely. Therefore, the microwave engineer will require some basic knowledge of practical issues involved in this connection, such as bringing cables into the building, terminating them, and interfacing (and cross-connecting) other types of equipment. The National Electrical Code (NEC) identifies three different intrabuilding regions with regard to cable placement.

- *Plenums.*   A plenum area is a compartment or chamber that forms part of the air distribution system and to which one or more air ducts are connected. A room with a primary function of air handling is also considered to be a plenum space.

- *Risers.*   A riser is an opening or shaft through which cable may pass vertically from floor to floor in a building.

- *General-purpose areas.*   These are other indoor areas that are not plenums or risers.

Cables are specifically listed for use in each of these areas. The NEC allows the use of a cable with a more stringent listing to be used in an application requiring a lesser listing, but not the other way around. When a cable run will be exposed to both indoor and outdoor environments, there are several options to consider. The first option is to run an outside plant cable for the entire run. The second option is to use an indoor/outdoor cable. The third option is to transition-splice the outside plant cable to an inside plant cable.

The consideration and design of proper cable pathways (i.e., conduit, cable trays, riser shafts, and so forth) and termination spaces (i.e., main/intermediate cross-connects, horizontal cross-connects, and the work area) are as important as the design of the cable network. For more information, see "Commercial Building Standard for Telecommunications Pathway and Spaces," TIA/EIA-569 or CSA-530. The primary focus of this standard is to provide design specifications and guidance for all building facilities relating to telecommunications cabling systems and components. Another useful document is the *Telecommunications Distribution Methods Manual,* available from the Building Industry Consulting Service International (BICSI).

6.4.1   Metallic Cables and Ground Potential Rise

Wireless operators are increasingly employing electrical utility facilities for the installation of their PCS and microwave antennas. It is important to understand that electric transmission towers, and electric utility environments in general, are dangerous, hazardous locations that are prone to a phenomenon called *ground potential rise (GPR)*. Communications cables (for example, E1/T1 circuits) are susceptible to damage from lightning surges, since they can develop high shield-to-pair voltages, even with low lightning currents on the shield; these voltages could cause lengthy and potentially catastrophic outages. GPR usually occurs during a power fault when the fault current returns to the power neutral source through the earth. Communications cables are considered to be exposed to GPR when the possibility exists that the local ground (at the cell-site) differs from remote ground by 300 V or more. Optical isolators (opto-couplers) are generally used to treat each circuit (including T1/E1 circuits) going into the power station. This protects the circuits and facility, as well as personnel, from electrical hazards associated with GPR. Design of any communication circuits into the power location has to be in accordance with ANSI/IEEE 367, "Recommended Practice for determining the Electric Power Station Ground Potential Rise and Induced Voltage From a Power Fault," and ANSI/IEEE 487, "Guide for Protection of Wireline Communications Facilities Serving Electric Power Stations."

The only alternative to GPR study would be a letter from the power utility owner, on letterhead and appropriately signed, stating that at no time will the GPR at the site or sites ever exceed 1,000-V peak-asymmetrical, as calculated per ANSI/IEEE 367. The power utility owner must be aware that issuance of the letter constitutes an assumption of liability for injury and/or damage brought about by electrical fault conditions.

6.4.2   Fiber-Optic Cables

Every fiber-optic cable requires proper installation techniques. Building codes and standards, environmental issues, proper design, routing, installation equipment, topologies, applications, and reliability concerns all have to be addressed. Considerations of tensile strength, ruggedness, durability, flexibility, size, resistance to the environment, flammability, temperature range, and appearance are also important in constructing optical fiber cable.

From the outside, a fiber-optic cable looks like any other electrical multiconductor cable. However, it is lightweight and flexible as compared to metal conductor cable. Typical fiber cable outside dimensions (ODs) range from less than 1/8 in up to 3/4 in, depending on fiber numbers and cable construction. The most common fiber-optic cable jacket materials are polyethylene (all types), PVC, and polyurethane.

Probably the most common mistake made by inexperienced fiber installers is to violate the minimum bending radius by making tight bends in the cable. Tight bends, kinks, knots, and other flaws in fiber cable can result in a loss of performance. The minimum bending radius in traditional fiber cable is usually in the range of 20 times cable OD—considerably higher than for electrical cable. However, new fiber technologies are lowering this minimum bend radius. The specific minimum bending radius for a particular cable should be researched in the cable manufacturer's specifications.

6.4.3 Digital Signal Cross-Connects

Microwave system is usually a part of a bigger transmission network and the digital signal leaving microwave radio and multiplexing equipment has to be terminated at the cross-connect point. *Digital signal cross-connect (DSX)* devices (also called *jackfields*) are used to connect one piece of digital telecommunication equipment to another. They simplify equipment connections and provide convenient test access and tremendous flexibility for rearranging and restoring circuits. DSX panels are available in numerous configurations and sizes to fit a wide variety of applications, but they all perform the same function. They terminate equipment and provide temporary jack access for centralized testing, cross-connection, reconfiguration, and restoration of various digital circuits. They are used in voice, data, and video networks for patching during equipment installations or breakdowns, network expansion, or traffic-pattern adjustments.

At T1 (1.544 Mbps) or E1 (2.048 Mbps) rates, DSX-1 panels connect network equipment such as office repeater bays, channel banks, multiplexers, digital switches, microwave radios, and so on. At the T3 (44.736 Mbps) rate, DSX-3 panels provide terminations for the high-speed (DS3 rate) side of the M13 multiplexers, and the low-speed (DS3 rate) side of the digital microwave radio and fiber-optic systems. The DSX-3 supports networks operating at the DS3 rate and the STS-1 and STS-3 electrical SONET rates. It is usually placed between network elements such as fiber-optic terminals, multiplexers, broadband digital switches, and digital cross-connect systems, and digital radio components, in a central office, CEV, hut, or cabinet, or on the customer's premises.

A *connecting block* (also called a *terminal block,* a *punch-down block,* a *quick-connect block,* or a *cross-connect block*) is a plastic block containing metal wiring terminals used to establish connections from one group of wires to another. Usually, each wire can be connected to several other wires in a bus or common arrangement. There are several types of connecting blocks: M66 clip, BIX, Krone, 110, and others. A connecting block has insulation displacement connections (IDCs), which means that it is not necessary to remove the insulation from around the wire conductor before "punching it down" (i.e., terminating it). Blocks can be reused for at least 500 terminations with less than 1 mΩ of connection resistance.

6.5 Microwave Antennas, Radomes, and Transmission Lines

The microwave antenna is a highly directional, parabolic, dish-shaped radiator that is connected to the microwave transmitter (through the RF branching network—RF circulators, filters, couplers, and switches) via transmission lines (coaxial cable or waveguide). Initial design and experimentation with microwave antennas began more than 100 years ago, and microwave system work using parabolic antennas grew significantly during the 1930s. During World War II, designs such as pencil-beam and shaped-beam antennas were developed for radar systems used by the allies. While many advances were made at this time, it was in the 1950s that terrestrial microwave communication systems were deployed, and parabolic reflector designs were utilized on these commercial systems.

Microwave antennas are available in many sizes to satisfy the requirements of a particular application. Generally speaking, the larger the antenna diameter, the higher the antenna gain relative to isotropic antenna and the smaller the beamwidth.

There are three basic types of *waveguides*: rigid rectangular, rigid circular and semiflexible elliptical. Short sections of flexible waveguide are also used for connections between antennas and the microwave radio. In all cases, it is desirable to keep the number and length of flexible sections as small as possible, since they tend to have higher losses and poorer VSWR than the main waveguide types.

Semiflexible elliptical waveguides are the most common type used for terrestrial microwave systems today. They are precision formed from corrugated high-conductivity copper and have an elliptical cross-section. The corrugated wall gives the waveguide increased crush strength, light weight, and relative ease of handling.

Coaxial cable is used in case of the split configuration (indoor/outdoor units) of the microwave radio. Coax cables are much easier to install, and today they are used more often than waveguides.

There will always be some loss of RF signal strength through the cables and connectors used to connect to the antenna. This loss is directly proportional to the length of the cable and inversely proportional to the diameter of the cable. Additional loss occurs for each connector used, and this must be considered in planning. Cable vendors usually can provide a chart indicating the loss for various types and lengths of cable.

6.5.1 Basic Antenna Parameters

A feed system is placed with its phase center at the focus of the parabola. Ideally, all the energy radiated by the feed will be intercepted by the parabola and reflected in the desired direction. To achieve maximum gain, this energy should be distributed such that the field distribution over the aperture is uniform. Because the feed is small, however, such control over the feed radiation is unattainable in practice. Some of the energy actually misses the reflecting area and is lost; this is commonly referred to as *spillover*. In addition, the field is generally not uniform over the aperture but is tapered, with maximum signal at the center of the reflector and less signal at the edges. This *taper loss* reduces gain, but the field taper provides reduced sidelobe levels.

A common way to define a parabolic dish shape is with the F/D ratio, where F is the focal length, and D is the diameter of the dish; the smaller the ratio, the "deeper" the dish. Most commercial microwave antennas use an F/D ratio of 0.25 to 0.38, with 0.32 to 0.36 the most common. The F/D ratio for a reflector can be determined by measuring the depth of the dish from the plane of the rim to the vertex at the center and using the basic equation for a parabolic curve. Typically, only the measurement from the vertex to the rim is required, since a parabola of revolution consists of the same shape curve for all radial sections.

During the 1980s, the need became greater for a lower-profile microwave antenna that also exhibited superior pattern performance. Two forces drove this requirement. One was the need to reduce the visual effect of radio communication installations. The other was the need to place more and more microwave "links" in the same geographic area.

6.5.1.1 Antenna gain. The gain of any antenna is essentially a specification that quantifies how well that antenna is able to direct the radiated radio frequency (RF) energy in a particular direction. Thus, high-gain antennas direct their energy more narrowly and precisely, and low-gain ones direct energy more broadly. According to the *reciprocity theorem,* the transmitting and receiving patterns of an antenna are identical at a given wavelength.

An antenna's gain and pattern are fundamentally related; indeed, they are really the same thing. Higher-gain antennas always have narrower beamwidth (patterns), and low-gain antennas always have wider beamwidth. For a parabolic reflector microwave antenna (above 1 GHz) consisting of a dish-shaped surface illuminated by a feed horn mounted at the focus of the reflector, gain is given as

$$\text{Antenna gain (dBi)} = 20 \log_{10}(D) + 20 \log_{10}(f) + 7.5 \text{ (feet)}$$

or

$$\text{Antenna gain (dBi)} = 20 \log_{10}(D) + 20 \log_{10}(f) + 17.82 \text{ (meters)}$$

where dBi = decibels over an isotropic radiator
       D = antenna dish diameter in feet or meters
       f = frequency in GHz

The above formula is based on the efficiency of a parabolic antenna being around 55 percent. Some manufacturers may be able to improve on this value, so the gain given by a manufacturer for a specific antenna should be used when available. In other cases, the above formula will suffice.

Engineers commonly refer to half-wave dipole antenna gains. Compared to the gain of an "ideal" isotropic antenna of 1 (0 dB), the gain of a half-wave dipole antenna is 1.64 (2.15 dB). The relationship between the two units, dBi and dBd, is therefore expressed as follows:

$$X \text{ (dBi)} = Y \text{ (dBd)} + 2.15$$

Considering both stations of a radio link, the difference between free-space loss comparison using isotropic and half-wave dipole antennas is about 4.30 dB.

Antenna gain and other specifications are defined and measured in the far field of the antenna. In the near field, antenna gain can be up to 10 dB lower than in the far field, and its value is used only for some very specific applications; for example, a back-to-back repeater in which one hop of the system can be only a few hundred feet long.

6.5.1.2 Half-power beamwidth. The half-power beamwidth is the angular separation between the half-power points on the main beam antenna radiation pattern, where the gain is one-half the maximum value (−3 dB).

Antenna gain and beamwidth are interrelated quantities and are inversely proportional; thus, the higher the gain an antenna has, the smaller the beamwidth. Therefore, increased care must be taken when aligning high-gain antennas to ensure that the antenna is accurately aligned on the center of the main beam—which could be only a few degrees wide.

The beamwidth of a parabolic antenna can be approximated by the following formula:

$$BW = 70/(f \times D)$$

where f = frequency (GHz)
$\quad\quad D$ = diameter of parabola (feet)

6.5.1.3 Polarization. Polarization is a physical phenomenon of radio signal propagation, and the term refers to the orientation of the electric field vector in the radiated wave. For linear polarization (horizontal or vertical), the vector remains in one plane as the wave propagates through space. There are two categories or types of polarization: linear and circular. Each has two subcategories: vertical or horizontal for linear, and right- or left-handed for circular. In general, any two antennas that form a link with each other must be set for the same polarization. This is typically accomplished by the way the antenna (or just the feedhorn) is mounted and as such is almost always adjustable at (or after) the time of antenna installation.

It is difficult to predict the orientation of the electric field in the near-field region, as the transmitting antenna cannot be considered as a point source in this region. In the far-field region, the antenna becomes a point source, the electric and magnetic components of the field become orthogonal to the direction of propagation, and their polarization characteristics do not vary with distance.

Note that, if the physical waveguide connection at the antenna is vertically oriented, the antenna has horizontal polarization, and vice versa.

For licensed links, the polarization may be specifically dictated by the terms of the license; for unlicensed links, the operator is free to choose, and the choice may be crucial in averting or correcting an interference problem. Note that, for most microwave (dish) antennas, it is not possible to determine the exact type of polarization the antenna is set up for by observation from a distance (such as when viewing a tower-mounted antenna from the ground).

If two antennas both had linear polarization, but one had vertical polarization and the other had horizontal polarization, they would be *cross-polarized*. The term *cross-polarization* (or *cross-pol*) is also used to generally describe any two antennas with opposite polarization. Cross-polarization is sometimes beneficial. An example of this would be a situation in which the antennas of link A are cross-polarized to the antennas of link B, where links A and B are two different but nearby links that are not intended to communicate with each other. In this case, the fact that links A and B are cross-polarized is beneficial because the cross-polarization will prevent or reduce any possible interference between the links.

On over-water paths at frequencies above about 3 GHz, it is advantageous to choose vertical polarization over horizontal polarization. At grazing angles greater than about 0.7°, a reduction in the surface reflection of 2 to 17 dB can be expected over that at horizontal polarization.

6.5.1.4 Radiation patterns. Radiation patterns describe the distribution in space of electromagnetic energy generated by a given antenna. The patterns are presented as polar plots (relative energy level versus angular position) in the E-plane, and H-plane—in other words, in the same plane as the E-field and H-field, respectively.

A simplified version, called a *radiation pattern envelope (RPE)*, is often used for design purposes. In this case, the pattern is deliberately "linearized," and the normal (sometimes wide) fluctuations in the field are removed. Radiation patterns (or the RPE version) are used to design and evaluate system performance as it relates to transmission (EIRP) in any given direction, or reception (RSL) from any given direction, including interference.

Four RPEs are always needed when analyzing the behavior of microwave antennas: two parallel envelopes (named HH and VV) and two crossed envelopes (named HV and VH). These acronyms mean

- HH. The response of a horizontally polarized antenna port (H-port) to a horizontally polarized microwave radio signal

- VV. The response of a vertically polarized antenna port (V-port) to a vertically polarized microwave radio signal

- HV. The response of a horizontally polarized antenna port to a vertically polarized microwave radio signal

- VH. The response of a vertically polarized antenna port to a horizontally polarized microwave radio signal

In recent years, significant advancements in the antenna-design areas of polarization discrimination and sidelobe reduction have provided the capability of enhanced spectrum efficiency in point-to-point microwave radio communications.

6.5.2 Microwave Antenna Selection

Standard, open-grid (for low wind loading), and high-performance dishes, and single or dual polarized antenna models, are available. Many antennas come with galvanized steel mounts based on EIA standards RS195B and RS222C. As the antenna beamwidth decreases, antenna alignment (and thus stability) become more critical. Furthermore, weight and wind loading are greater with large antennas. As a consequence, the antenna mounting structure must be several times more rigid (against twist and sway) for each successive increase in antenna size. The selection of antenna size should be based on the results of path analysis and calculations. The antenna size must be determined before a frequency coordination study can be performed and before applying for the license to operate the microwave system.[5]

Main parameters of interest when choosing the microwave antenna are as follows:

1. Operating frequency band

2. Radiation pattern

3. Gain

4. Polarization (single or dual polarized)

5. Half-power beamwidth

6. Wind load

7. Front-to-back ratio

8. Cross-polarization discrimination (dB) (This is the difference between the peak of the co-polarized main beam and the maximum cross-polarized signal over an angle twice the 3-dB beamwidth of the co-polarized main beam)

9. Isolation (between inputs of single-band, dual polarized antennas)

10. Additional options for the most microwave antennas that may be ordered from manufacturers are

   – Input flanges

   – Antenna color variations

   – Radomes (reflector protectors) of various colors

   – High-wind survival options

   – Corrosive environment protection

   – Packing type and quantity options

   – Various reflector types

   – Special-purpose antennas

   – High-performance or shrouded antennas

   – Special accessories such as struts, ice shields, and so on

Optional high-performance antenna versions include an RF shroud to improve sidelobe performance and a planar radome to protect the antenna against ice or snow accumulation. The amount of loss for a radome may vary from less than 0.5 dB for a typical unheated radome to more than 2.0 dB for a typical heated radome in the high-frequency bands.

During a rainstorm, water sheeting on the face of the antenna causes an additional loss called *wet radome loss*. Preliminary analyses suggest that the effect of water on the antenna radome may be an important mechanism, in addition to rain attenuation along the path.[6] This is particularly true in tropical regions where the rainfall rate is very high, and microwave links operating in the higher microwave bands will suffer from high attenuations due to rain. The losses of a wet radome depend not only on the rainfall rate but also on wind conditions during the rain.

The loss for a Teflon®-coated radome is 1 dB, whereas the loss for a fiberglass or ABS plastic radome depends on the rain rate and frequency band but may be much higher than the Teflon radome. The Hypalon® radome material used on some high-performance dishes can be as bad as fiberglass. Because of the lower wet-radome losses, Teflon-coated radomes are recommended for bands that are affected by rain outage. Teflon is also used on most high-performance antennas.

6.5.3 Microwave Antenna Installation

The last exercise during the microwave radio-hop installation is to align the antennas. The main purpose of path alignment is to physically align the antenna's azimuth and elevation for maximum signal transfer (minimum path loss). Optimal *antenna alignment* occurs when both transmitting and receiving antennas are precisely aimed at each other in both azimuth and elevation. Azimuth is the angle in the horizontal plane with respect to true north, and elevation is the angle in the vertical plane with respect to the horizontal plane.

Antenna altitude and *antenna height* are expressed in meters or feet. These distances are measured to the center (feeder) of the parabolic dish antenna.

It is important that the antennas be properly aligned, meaning that the maximum radiations of the two antennas are directly pointed toward each other. It is also important to ensure that the two antennas for the link are not cross-polarized and that each antenna is pointed or aligned to maximize the received signal level. A coarse alignment can be done by using line of sight or a compass. The antenna is fine-aligned by adjusting for maximum input power using the outdoor unit's audible signal or AGC voltage. At one end of the link at a time, the antenna pointing direction is carefully adjusted to maximize (or *peak*) the reading on the indicator tool. Once the maximum signal is achieved, the antennas are aligned for elevation optimization.

After performing this procedure on both ends, it is very important to obtain the actual received signal level in dBm so as to verify that it is within 0 to 3 dB of the value obtained from the link budget calculation. If the measured and calculated values differ by more than about 8 dB, it is possible that either the antenna alignment is still not correct or that there is another problem in the antenna/transmission line system—or both.

It is possible to get a *peak* reading during the antenna alignment process if one or both of the antennas are aligned on a *sidelobe*, in which case the measured receive level may be 20 dB (or more!) lower than the calculated value indicates that it should be (Fig. 6.3).

Traditionally, the radios that will be placed at each site are used to complete the task of optimizing the path. However, there are several reasons for not utilizing the radios to complete the process. The radios may not be available at the time the test has been scheduled, or their reliability may be questionable, thus requiring alternative methods. Another possible situation when the radios might not be usable is when the FCC permits have not been granted, but the contractor needs to complete the path test on time to meet the customer's requirements or to stay ahead of expected turbulent weather. In addition, if the anticipated path is questionable, a quick, cost-effective, reliable method is needed to test the link prior to the significant investment in constructing towers, purchasing radios, and obtaining other expensive equipment and hardware.

Alternative test instrumentation must be utilized, in lieu of the radios, whenever the previously mentioned circumstances arise. Scheduling of the path alignment test and installation of the associated hardware (e.g., cables, waveguide, antenna) can be facilitated to reduce excess mobilization costs. Some of the most widely used appara-

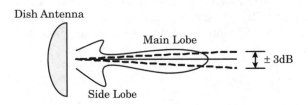

Figure 6.3 Radiation pattern of a directional antenna.

tus for this application are signal generators (used as the transmitter) and spectrum analyzers (used as the receiver). The signal generator should be a broadband, synthesized device (phase locked to a reference clock) with accurate output power, equal to or greater than 0 dBm. The spectrum analyzer should be tunable, and have at least −100 dBm of sensitivity at the frequency band of interest.

One of the most unpredictable and unacceptable unavailability events in the microwave system is power fading due to the ducting, antenna decoupling, and obstruction in coastal and similar humid climates. One of the simplest solutions is to uptilt larger dishes on longer paths (at above 6 GHz) in ducting areas perhaps 1 dB to minimize power fading and multipath activity due to the nighttime antenna decoupling.

In space-diversity microwave systems, *diversity antenna delay equalization (DADE)* has to be set in receivers to ensure errorless or hitless data switching in a severe fading environment. This requirement is a result of different waveguide lengths from the microwave radio to the main and diversity antennas. On most MW radios today, the DADE is automatically adjusted based on the phase difference between main and space diversity received signals.

Any system components mounted outdoors will be subject to the effects of wind, and it is important to know the direction and velocity of wind that is common to the site. Antennas and their supporting structures must be able to prevent these forces from affecting the antenna or causing damage to the building or tower on which the components are mounted. Antenna designs react differently to wind forces, depending on the area exposed to the wind *(wind loading)*. Most antenna manufacturers will specify wind loading for each type of antenna manufactured.

For reasons of safety it is important to know that the near-field regions can constitute radiation hazards at substantial physical distances from an antenna, especially if the antenna is large ($D \gg \lambda$).

When planning microwave paths longer than seven miles, the curvature of the earth becomes a factor in path planning and requires that the antenna be located higher off the ground. The *minimum antenna height* at each end of the link for paths longer than seven miles (assuming smooth terrain without obstructions) is the height of the first Fresnel zone plus the additional height required to clear the Earth's bulge.

The approximate formula is

$$h = 43.3 \sqrt{\left(\frac{d}{4f}\right)} + \frac{d^2}{8}$$

where h = height of the antenna (feet)

d = distance between antennas (miles)

f = frequency in GHz

6.5.4 Transmission Line Installation

Elliptical waveguide installation is a very tricky job that must be performed by trained and experienced technicians. Very small deformations can introduce enough impedance mismatches to produce serious problems.

Elliptical waveguide may be ordered with the connectors attached and tuned if the precise length is known in advance. If the connectors are to be installed in the field, it may be necessary to sweep and tune the waveguide and connectors to realize optimal return loss. The waveguide must be supported with special hangers at regular intervals to prevent stress, movement, and excessive pulling forces caused by the cumulative weight of the waveguide. Hangers are usually installed at 3-ft (1-m) intervals. Waveguide (as well as coaxial cable) must be grounded using special grounding kits. It is common practice to ground the top of the waveguide run, the bottom of the run, and at the radio terminal.

As we mentioned earlier, coaxial cable is used in case of a split configuration (indoor and outdoor units) of the microwave radio. Coax cables are much easier to install, and they are presently used more often than waveguides. Some mechanical characteristics of the coaxial cable interconnecting IDU and ODU or waveguide are also important. These include dimensions over the jacket or outer diameter, distributed weight, and minimum bending radius. For waveguides making the connection from the indoor branching to the outdoor antenna, the important specification is the minimum bending radius (E-plane and H-plane).

For coaxial cables interconnecting IDUs and ODUs operating in intermediate frequency (IF) or baseband (BB), the maximum acceptable length is vital information for the installation. Usually, the maximum length is about 300 m (984 ft) because of the electrical limitations of the coaxial cable.

Generally speaking, when the environment is hazardous because of the presence of chemicals, explosive atmospheres, corrosive environments, or shock hazards, or when there are many T1/E1 signals traveling across the line (such as STM-0 or STM-1), it is a good approach to provide physical protection and/or separation. The duct can also act as an obstacle against vandalism, the influence of ultraviolet rays, water penetration, and fire, thus preserving the cable(s) inside it.

The waveguide and air dielectric coaxial cable must be pressurized with equipment designed to dry the pressurized air. This action is necessary to displace any water vapor from the waveguide or cable that will significantly increase attenuation. The pressurization equipment may be either a mechanical compressor or a container of compressed gas (which must be replaced regularly) equipped with a pressure regulator. Several types of mechanical compressors are available with various capacities, operating from AC or DC line voltage and operating in manual or automatic drying cycles. Pressurization equipment is normally attached to the indoor end of the waveguide or coax cable, through a manifold.

6.5.5 Installation Safety and Security Issues

The installation team must have enough people skilled in the different aspects of the microwave project (i.e., installers, riggers, and commissioners) to ensure that the project is completed in a professional manner and on schedule. It is obvious that communications tower safety is a life-and-death issue not only for the men and women who work on the structures but also for the companies that own them. Despite the common practice of contracting out the construction and maintenance of the towers to other companies, the owners of the structure (company directors, officers, and shareholders) share a legal responsibility with the subcontractor to make sure the work is carried out safely. This situation is not a new one—most jurisdictions require property owners to provide safe job sites regardless of whether they directly employ the workers.

Communications towers are very high and very narrow and can be dangerous. Therefore, tower construction and microwave antenna installation require special skills (Fig. 6.4). Without the cranes to lower pieces into place from above, the towers are constructed with an appa-

Figure 6.4 Microwave antenna installation.

ratus known as a *gin pole*. The gin pole is used to hoist material from the ground and past the worker at the top so that it can be positioned and bolted in place. The work is carried out at the mercy of wind and inclement weather at heights that reach up to few hundred meters. Typically, communications towers range between 30 and 150 m (100 to 500 ft) in height.

While working on a communications tower, very stringent work procedures and reliable safety equipment are required to protect workers from falls. A subcontractor must carry sufficient liability insurance for services provided within or outside North America, and the copy of the insurance coverage document has to be provided to the project manager prior to the start of any installation.

6.5.6   Return Loss Measurements

Voltage standing wave ratio (VSWR) is a measure of impedance mismatch between the transmission line and its load. The higher the VSWR, the greater the mismatch. The minimum VSWR (i.e., that which corresponds to a perfect impedance match) is unity. The combination of the original wave traveling down the coaxial cable (toward the antenna, or the opposite during receive) and the reflecting wave is called a *standing wave*. The ratio of the two above-described waves is known as the *standing wave ratio. Return loss* is basically the same thing as VSWR. If the antenna absorbs 50 percent of the signal, and 50 percent is reflected back, we say that the return loss is −3 dB. An antenna might have a value of −10 dB (90 percent absorbed and 10 percent reflected).

Impedance, in antenna terms, refers to the ratio of the voltage to current (both are present on an antenna) at any particular point of the antenna. This ratio of voltage to current varies on different parts of the antenna, which means that the impedance would be different on different spots on the antenna if you could pick any spot and measure it. The impedance for the entire chain from the radio to the antenna must be the same, and almost all radio equipment is built for an impedance of 50 Ω.

One common way of visualizing the VSWR is to use a polar plot called a Smith chart. From this plot for the VSWR value, the return loss and the impedance for the different frequencies can be derived. Therefore, it is an important instrument for understanding antennas and transmission lines.

VSWR/return loss can be measured for the antenna, transmission line, transmission line and all rigid components, and for the entire antenna subsystem (e.g., antenna, transmission line, connectors, and so on). These measurements ensure correct impedance matching between the radio port and antenna port. In addition, the insertion loss of the transmission line and all other components should be measured.

Measurement of the antenna subsystem should be performed as follows:

1. The radio should be switched off before testing the waveguide.

2. A directional coupler should be installed between the radio Tx port and the transmission line, and polarity should be checked, as well as flange connections

3. The power meter and power sensor are then connected to the measurement port on the directional coupler.

4. The radio power should now be switched on.

5. The reflected power should be measured taking into account the coupling coefficient inside the directional coupler.

6. The reflected power should be around 30 dB below the measured radio output power measured for the system gain calculation.

7. The following values should then be recorded:

   – Transmit power (dBm)

   – Directional coupler coupling coefficient (dB)

   – Power meter reading (dBm)

   – Measured reflected power (dBm)

   – Return loss (dB)

8. A plot of the sweep across the frequency band should be provided as well. using a reference of 0 dBm.

Return loss measurement of just the transmission line may be performed by placing an ideal termination at the opposite end of the transmission line. Insertion loss may be measured by shorting the transmission line. A shorting connector would be used for coaxial cable, and a shorting plate would be used for the elliptical waveguide.

6.6 GIS Data

6.6.1 Datums and Geometric Earth Models

When producing maps, the surface of the Earth has to be mapped onto a plane. This is called a *projection*. A projection from a curved surface like the Earth to a plane cannot be done without distortion; therefore, various projection types with different advantages and disadvantages have been developed and are still in use throughout the world. Every map in the world has been made with one or another of such projections, e.g., UTM, Gauss Kruger, or Lambert Conformal. The same area mapped by two different projections may look slightly or completely different.

A *spheroid* is a surface obtained by rotating an ellipse around one of its axes. In geography, the shape of the Earth is considered to be a spheroid. However, the Earth is not a perfect spheroid; therefore, several spheroid models have been adopted to approximate the surface in different parts of the Earth. These spheroids are slightly different in shape and size. They are uniquely defined by their equatorial and polar radii, and they are identified by names like Clarke 1866, Hayford, and others.

Every map is associated with a projection and a spheroid. This is called a *geodetic datum* (from Latin, singular for data-given things). Hundreds of different datums have been used to frame position descriptions since the first estimates of the Earth's size were made by Aristotle. Datums have evolved from those describing a spherical Earth to ellipsoidal models derived from years of satellite measurements. Modern geodetic datums range from flat-Earth models used for plane surveying to complex models used for international applications that completely describe the size, shape, orientation, gravity field, and angular velocity of the Earth.

Referencing geodetic coordinates to the wrong datum can result in position errors of hundreds of meters.

The accuracy of a geographic database depends on the source material, data capture technique, and resolution. The source material can be paper maps, aerial photography, or satellite images. Examples of paper maps with different scales and accuracy are topographical, ordnance survey, and city maps.

Different nations and agencies use different datums as the basis for coordinate systems used to identify positions in geographic information systems, precise positioning systems, and navigation systems. The diversity of datums in use today and the technological advancements that have made possible global positioning measurements with submeter accuracies require careful datum selection and careful conversion between coordinates in different datums.

True geodetic datums were employed only after the late 1700s, when measurements showed that the Earth was ellipsoidal in shape. In North America, there is, for example, NAD83 (an acronym for North American Datum 1983). Sometimes still used is NAD27 (North American Datum 1927). NADCON is the United States Federal standard for NAD27 to NAD83 datum transformations.

The GPS is based on the World Geodetic System 1984 (WGS-84). There are numerous other horizontal data, such as Yacare (South America), Tokyo (Japan), Djakarta (Indonesia), Easter Island (Pacific Ocean). The right choice of the horizontal datum is crucial for good path calculations. Coordinate values resulting from interpreting latitude, longitude, and height values based on one datum as though they were based in another datum can cause position errors in three dimensions of more than a kilometer.

6.6.2 Coordinate Systems

Coordinate systems to specify locations on the surface of the Earth have been used for centuries. In western geodesy the equator, the tropics of Cancer and Capricorn, and then lines of latitude and longitude, were used to locate positions on the Earth. Eastern cartographers used other rectangular grid systems as early as A.D. 270, and various units of length and angular distance have been used over history. The meter is related to both linear and angular distance, having been defined in the late 18th century as one-ten-millionth of the distance from the pole to the equator.

The most commonly used coordinate system today is the *latitude, longitude, and altitude system*. The prime meridian and the equator are the reference planes used to define latitude and longitude. The *geodetic latitude* (there are many other defined latitudes) of a point is the angle from the equatorial plane to the vertical direction of a line normal to the reference ellipsoid. In other words, latitude measures how far north or south of the equator a place is located. The equator is situated at 0°, the north pole at 90° north (or simply 90°, since a positive latitude number implies north), and the south pole at 90° south (or −90°). Latitude measurements range from 0° to (±) 90°.

The *geodetic longitude* of a point is the angle between a reference plane and a plane passing through the point, both planes being perpendicular to the equatorial plane. Longitude measures how far east or west of the prime meridian a place is located. The prime meridian runs through Greenwich, England. Longitude is measured in terms of east, implied by a positive number, or west, implied by a negative number. Longitude measurements range from 0° to (±) 180°.

The geographical coordinates (latitude and longitude) are typically expressed in the sexagesimal form (i.e., degrees, minutes, and seconds). Sometimes the coordinates can be shown in *universal transverse mercator (UTM)* format or in values expressing north or east, for example. When the positioning coordinates are obtained from topographical charts, they are normally expressed in UTM format.

The geodetic height at a point is the distance from the reference ellipsoid to the point in a direction normal to the ellipsoid. Altitude is measured with reference to mean sea level (MSL), and height is measured with reference to the soil or ground level. In this application, altitudes are known as *above mean sea level (AMSL)* and heights as *above ground level (AGL)*. It is essential to show the vertical or altimetric reference datum (or VRD) with the altitude, because altitudes change when the vertical datum changes.

6.6.3 The Global Positioning Systems

The *global positioning system (GPS)* is based on information users receive from satellites. The purpose of GPS is to provide users with the ability to compute their location in three-dimensional space.[7] To accomplish that, the receiver must be able to lock onto signals from at least four different satellites. Moreover, the receiver must maintain a lock on each satellite's signal long enough to receive the information encoded in the transmission. Achieving and maintaining a lock on four or more satellite signals can be impeded, because the signal is transmitted at 1.575 GHz, a frequency that is too high to bend around or pass through solid objects in the signal's path. This why GPS receivers cannot be used indoors. Outdoors, tall buildings, dense foliage, and terrain that stand between a GPS receiver and GPS satellite will block that satellite's signal.

In addition, the signals from the GPS satellites previously were degraded intentionally by the U.S. Department of Defense (DoD) for purpose of national security. This performance degradation is known as *selective availability (SA)*, and only DoD-approved users have access to satellite signals without SA. However, a policy statement issued by the White House indicates that SA was turned off in 2000, and the accuracy of GPS position fixes has been improved significantly.

6.6.4 DEM Data

A digital elevation model (DEM) is a digital file consisting of terrain elevations for ground positions at regularly spaced horizontal intervals. The United States Geological Survey (USGS) produces five different digital elevation products. Although all are identical in the manner the data are structured, each varies in sampling interval, geographic reference system, areas of coverage, and accuracy. The primary differing characteristic are the spacing, or sampling interval, of the data. The five versions of DEM are

- 7.5-minute DEM (30 × 30-m data spacing). Based on "1:24,000 scale" topographic maps

- 1-degree DEM (3 × 3-arc-second data spacing), also referred to as "3-arc second" or "1:250,000 scale" DEM data

- 2-arc-second DEM (2 × 2-arc-second data spacing)

- 15-minute Alaska DEM (2 × 3-arc-second data spacing)

- 7.5-minute Alaska DEM (1 × 2-arc-second data spacing)

DEMs may be used in the generation of three-dimensional graphics displaying terrain slope, aspect (direction of slope), and terrain profiles between selected points. A DEM file is organized into a series of three records, A, B, and C. The A record contains information defining the general characteristics of the DEM, including its name, boundaries, units of measurement, minimum and maximum elevations, number of B records, and projection parameters. Each B record consists of an elevation profile with associated header information, and the C record contains accuracy data. Each file contains a single A and C record, while there is a separate B record for each elevation profile.

Elevation data from cartographic sources are collected from any map series 7.5 minute through 1 degree (1:24,000 scale through 1:250,000 scale). The topographic features (contours, drain lines, ridge lines, lakes, and spot elevations) are first digitized and then processed into the required matrix form and interval spacing. The digital elevation models distributed within the Department of Defense cover 1 × 1-degree blocks and are *called digital terrain elevation data level 1 (DTED-1)*. The DMA 1-degree DTED-1 data and USGS-distributed 1-degree DEMs are gridded by using the World Geodetic System 1984 (WGS 84).

Elevations are in meters relative to National Geodetic Vertical Datum of 1929 (NGVD 29) in the continental U.S. and local mean sea level in Hawaii. The 1-degree DEM mosaic data set is characteristically the same as the source 1 × 1-degree DEM unit of coverage.

6.6.5 Magnetic and True North

Many people are surprised to learn that a magnetic compass does not normally point to true north. In fact, over most of the Earth, it points at some angle east or west of true (geographic) north. The direction in which the compass needle points is referred to as *magnetic north,* and the angle between magnetic north and the true north direction is called *magnetic declination.* The terms *variation, magnetic variation,* and *compass variation* are often used in place of *magnetic declination.*

The magnetic declination does not remain constant in time. Complex fluid motion in the outer core of the Earth (the molten metallic region that lies from 2,800 to 5,000 km below the Earth's surface) causes the magnetic field to change slowly with time. This change is known to as *secular variation*. Because of secular variation, declination values shown on old topographic, marine, and aeronautical charts need to be updated if they are to be used without large errors. Unfortunately, the annual change corrections given on most of these maps cannot be applied reliably if the maps are more than a few years old, since the secular variation also changes with time in an unpredictable manner.

The current year's declination is oftentimes marked in degrees on the map, which is one reason why it is important to have a current map. Magnetic field models are used to calculate magnetic declination by means of computer programs such as the Magnetic Information Retrieval Program (MIRP), a software package developed by the Geomagnetism Program of the Geological Survey of Canada. The user inputs the year, latitude, and longitude, and MIRP calculates the declination. MIRP is able to compute values for any location on the Earth in the time period 1960 to 2000. It is possible to use the following web page to calculate declination by using MIRP:

http://www.geolab.nrcan.gc.ca/geomag/mirp_e.shtml

Navigating by compass requires determining bearings with respect to true or grid north from a map sheet and converting them to magnetic bearings for use with a compass. Important information on the map (see Fig. 6.5) is the magnetic declination diagram shown in the margin toward the left of the map. This will have a star with a straight line representing the North Star (true north) and the line (MN) to the right of it or left of it, depending on the area, to denote the magnetic declination that *must be applied* for compass correction to use the map for true headings. Another line with GN will also appear, which represents the *geographic (grid) north*.

- To change magnetic to true headings, we will add east declination and subtract west declination.

- To convert from true to magnetic headings, we will subtract east and add west declination.

It is important to distinguish between grid north, the direction of reference lines shown on the map, and true north. Although declination always refers to the angle between magnetic north and true north, it is often broken into two parts for convenience.

In microwave engineering documents, antenna orientation is always given relative to the true north, so it has to be adjusted during the site and path surveys and during equipment installation and alignment.

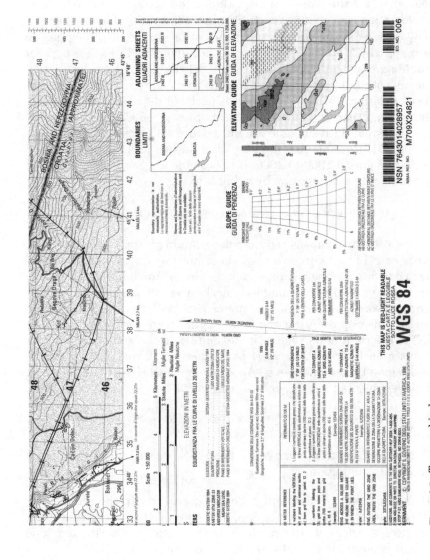

Figure 6.5 Topographical map.

6.7 Field Surveys

6.7.1 Site Surveys

In a first approximation, topographical and digital maps are sufficient to eliminate certain sites if, due to the terrain obstacles, they do not have a LOS with other sites. The second step would be a site visit and establishment of LOS availability for shorter microwave systems. For longer microwave hops, a detailed path survey will be required. General information includes location data and site details that the engineer must be aware of when designing a microwave system. The engineer should

- Note site details such as address, site directions, and access.
- Determine the *site type,* which identifies whether this is a rooftop shelter, cabinet or leasehold, or tower.
- Confirm that a suitable power (AC and/or DC) source exists for each radio unit.
- Confirm that each location has a suitable earth ground to support the installation of a lightning arrestor.
- Address other installation issues, such as building access and permits required by state or local governments. If the equipment is collocated with other users, an additional form with the relevant site and equipment information should be filled out.
- Ensure that there is available space for the radio equipment and antennas.

Any written observations and comments, such as potential obstacles, buildings, type of surrounding trees and vegetation, power lines and towers, water towers, lakes, rivers, airports, and communication towers, are very important to the microwave designer. Even more beneficial to the engineer are digital photographs that show possible antenna mount locations and the actual LOS paths as seen from the site. Panoramic pictures at every 30°, pictures of obstructions and existing facilities (e.g., shelter, tower), electrical connections, and Telco installations are recommended and can be very useful.

For tower sites, all relevant information must be provided, such as the condition of the tower and grounding (if it is an existing site), its capability to accommodate microwave antennas (e.g., twist and sway, available space, and so forth). Required loading needs to be calculated if new tower installations are proposed, and these must take into account the antenna wind and ice loading.

Cableways to the tower and in the tower should be documented with notes of available space and suggested clamps for the new cables. Take notes that show the dimensions of the cable ladders to determine what clamps to use.

The ease of service access for maintenance personnel, particularly tower-mounted equipment, can have a significant impact on costs and repair time. The site access should be described with notes regarding its condition and possible seasonal variations. Photocopied sections of maps with notes identifying the access from nearest highway are recommended.

6.7.2 Path Surveys

Selection of a suitable microwave radio site includes a number of factors. There are economical and engineering benefits to be gained by maximizing the sharing of infrastructure and sites between the various types of elements in the network, particularly regarding expensive civil infrastructure such as towers and equipment housings. Good microwave sites, particularly in relation to hub sites, will be relatively high points to provide the maximum line-of-site availability to other (spur) sites.

A path survey can be undertaken by visiting sites and observing that the path is clear of obstruction. It is important to make note of potential future interruptions to the path, such as tree or foliage growth, future building plans, nearby airports and related flight-path traffic, and any other transient traffic considerations.

Attention should be given to future network growth requirements in all areas, especially if the site is likely to develop into a future hub, as well as to informing landowners about it so as to prevent problems later. Attention should be paid to any local authority planning restrictions and approvals for structures or antenna installations planned.

A clear transmission path must exist between the two link nodes of any microwave radio link. Furthermore, as the radio wave disperses as it moves away from the source, there must exist additional clearance over and around any obstructions to prevent attenuation of the transmitted signal. This additional clearance, known as the *Fresnel zone,* differs for the frequency band of the radio path, where higher frequency translates into a smaller clearance requirement. Line of sight between two sites can be confirmed via map-based studies or direct survey. In either event, the surveyor must allow for future obstructions that may impinge the radio path. These can result from various causes, such as new buildings, tree growth, cranes, and so on. It is also important to notice whether there are some large bodies of water nearby, since large reflecting surfaces can produce problems as a result of increased multipath probability.

A *path profile* is a graphical representation of the path traveled by the radio waves between the two ends of a link. It is a result of path survey. The path profile determines the location and height of the antenna at each end of the link, and it ensures that the link is free of obstructions and propagation losses from radio phenomena such as multipath reflections. A path profile is established from topographical maps, which, by reference to the contours of the map, can be translated into an elevation profile of the land between the two sites in the path. Earth curvature can be added, as well as obstacles. The Fresnel zone calculation can then be applied and an indication of any clearance problems gained. Various software tools are available to assist this process if required, but most rely on the availability of topographical data to the appropriate degree of accuracy being available in digitized format.

For longer paths at lower frequencies and higher-capacity systems, it is imperative to actually perform physical path survey and to avoid relying on the maps or aerial photographs. Again, as microwave network planning is an iterative process, if line of sight cannot be achieved, this information should be processed back through the network plan and alternative path calculations and site selection performed.

6.8 Housing the Equipment

6.8.1 Shelters

Telecommunications equipment comes in all shapes and sizes. To properly protect that equipment, shelters are designed and often custom built to house it. Different materials can be used to construct the shelters, including metal frames, concrete, wood frames, and lightweight insulated materials. Each type of material offers benefits for particular environmental conditions and site locations. Wireless operators require shelters that precisely fit their needs, so shelter manufacturers must be flexible enough to design customized shelters. Strict zoning regulations, and the fact that most prime locations are already occupied, will require manufacturers to make shelters that can be positioned virtually anywhere or to use small cabinets for the telecom equipment infrastructure.

Shelters erected on the top of a 40-story building may need to withstand strong gusts of wind similar to those built for hurricane-prone areas. Bullet- and vandal-proof shelter and cabinet design are also quite often required; all shelters and cabinets must pass certain ballistics requirements as per Bellcore recommendations. To keep the equipment within a telecommunications shelter in proper working order, it is important to control temperature levels in the shelter. Temperature control units often are initially integrated into a shelter, but they can also be added to a structure to create the proper system conditions. In colder climates, the insulation is increased. If the shelter is placed beneath a tower in an area where ice is common, the building's construction material must be able to withstand ice falling from the tower. Common problems for many shelters are rainwater from leaking roofs and damp walls resulting from water condensation.

Grounding with a large buried ground loop, including many outward pointing arms, is probably the best method, but it can be difficult to build in densely urbanized areas. *It is important to remember to separate equipment grounding and protection grounding.*

Battery backup and/or diesel generators with extended-capacity diesel tanks are required in some areas. It is important to incorporate proper diesel fume exhaust and battery ventilation and also to remember that *all* batteries produce explosive hydrogen gas.

Access roads to the sites can be very costly. The general requirements for access roads is that they are to be permanent, with a minimum width of 3 m. The transportation of building material should be feasible without obtaining special permits for long vehicles or traffic police escorts. Prefabricated tower/mast sections consume a great deal of transportation space, but other constructions can be taken apart and packed in an economical manner.

6.8.2 Cabinets

Cabinets have been the time-proven components for housing telco equipment for many years and recently have become an option of choice for wireless and microwave equipment. They are relatively small, with efficient use of rack space for mounting the payload as well as the power plant.

They are exposed to the normal elements such as rain, snow, and temperature extremes that came with the seasons, so they must include several features to protect the equipment and batteries. Two essential features are *high- and low-temperature shutdown* and *temperature compensation.* The high/low temperature shutdown feature can work in two ways. First, the system can turn off the microwave and/or Telco equipment while still providing DC power to the cabinet fans or heaters, depending on whether the temperature is low or high. Second, the system can shut down completely if the cabinets use AC fans and heaters. Temperature compensation regulates the amount of battery voltage, depending on the battery temperature, thus prolonging the life of the batteries.

Remote monitoring also reduces costs (personnel labor and travel time) by enabling personnel to access the system without being physically present at the site. Remote monitoring via Ethernet connections allows operations to monitor vital data such as temperature, voltage, and alarms as well as capturing engineering data such as peak current draw.

6.8.3 Equipment Room

An equipment room is defined as any space where telecommunications equipment resides. In the design and location of the equipment room, one should provide space for expansion and consider water infiltration. Since the telecommunications equipment in this room is usually of large in size, delivery accessibility should be a consideration as well. The minimum recommended size for this room is 14 m^2 (150 ft^2). An equipment room may be a room in a building or a shelter that is placed either on the roof of a building or on the ground. The equipment room for the microwave equipment must fulfill a number of requirements. The building material used for walls, floor, and ceiling must be solid enough that it is possible to drill holes to fasten cable ladders to the wall or the ceiling and to fasten the cabinets to the floor. Floor space must be available for microwave equipment cabinets/racks, additional equipment, and future expansions. The room must have a lockable door that is large enough to allow for equipment transport. If the room has windows, it is recommended to cover them with blinds to minimize the heating effect of direct sunlight. The room must be clean and, preferably, have painted walls and ceiling as well as a painted floor or antistatic flooring to minimize dust.

For cabinet *leveling* purposes, the floor must be leveled to within ±3 mm/2000 mm, and the floor gradient must be within ±0.1. The floor must be able to carry the extra weight of the equipment or be reinforced.

The temperature in the equipment room must be kept within specified equipment limits. Heat generated by the equipment must be removed by ventilation or air conditioning.

Lighting protection must be adequate, and a power outlet for the connection of machine tools and test equipment must be available for installation and maintenance.

The *telecom cabinets* and *battery racks* are mounted on the floor. They may be positioned against a wall, back to back (in case of a small equipment room), or freestanding. Expansion cabinets and racks should be positioned to the right of the main cabinet (facing the cabinet) so as to follow the same standard layout globally. If the installation requirements include earthquake protection, the space between wall and cabinet is to be at least 100 mm, and between cabinets at least 150 mm.

Certain distances must be considered when planning the room layout to provide a convenient working environment during installation and maintenance work. A distance of 1000 mm (40 in) in front of the cabinets and racks work is recommended. Space for future expansion must also be considered. Make sure there is free space above the cabinets for exhaust air, and below the cable ladder to make it easier for bending antenna cables and waveguides. There must be enough free space above the ladder for maintenance work.

The minimum bending radius of cable (i.e., power, switchboard, coaxial, armored, fiber optics, ABAM, waveguides, and so on) **shall not** be less than the cable manufacturer's specification. Waveguides are typically run separately from all other cables, preferably in a separate raceway.

The location of the battery backup rack depends on the maximum allowable length of the battery cables. The acceptable power drop and cross-sectional area of the cables will limit the maximum length of the cables. Due to the heavy floor load, the battery rack is preferably placed in the corner of the room and/or on supporting beams.

It is important to determine what kind of safety equipment and alarms will be installed at a particular radio site. Fire extinguishers that are easily accessible, an emergency lighting system, and an alarm system are recommended at the radio site.

It is recommended that the alarm system at least support an intrusion alarm, a fire alarm, and high/low temperature alarms.

6.9 Microwave Antenna Mounting Structures

Many communities have very strict requirements and regulations concerning the siting of towers and antennas. Communities understand that they have to provide a reasonable opportunity for the siting of wireless telecommunications facilities, which enhances the ability of providers of wireless telecommunications services to provide such services to the community quickly, effectively, and efficiently. At the same time, they are sensitive to the effects on aesthetics, environmentally sensitive areas, historically significant locations, flight corridors, health and safety, and so on. They permit the construction of new towers only where all other reasonable opportunities have been exhausted, and they encourage the users of towers and antennas to configure them in a way that minimizes the adverse visual impact of the towers and antennas. Communities also require cooperation and collocation, to the highest extent possible, among competitors so as to reduce cumulatively negative consequences.

The construction standard for today's towers is primarily modeled after American National Standards Institute (ANSI) standards. Communication towers are designed in accordance with "Structural Standards for Steel Antenna Towers and Antenna Support Structures," ANSI/TIA/EIA-222-F, a nationally recognized standard jointly sponsored by the Electronic Industries Alliance (EIA, formerly the Electronic Industries Association) and the Telecommunications Industry Association (TIA). The ANSI standard is based on equations developed by professional engineers using wind tunnel testing to accurately predict the effect that wind has on telecommunication structures.

Most structures are designed to withstand a forceful wind speed that occurs on the average of once every 50 years. This wind speed is then escalated, with height, to a much higher wind speed at the top of the structure. A gust factor to account for the varying nature of wind is also incorporated into the design of the structure. The ANSI standard lists minimum 50-year return wind speeds for all counties within the United States. Radial ice accumulation is also accounted for in the design. In addition, safety factors specified by the American Institute of Steel Construction and the American Concrete Institute are used to design tower structures. The same factors used in building design are incorporated into both the structure and foundation design.

6.9.1 Monopoles, Self-Support Towers, and Guyed Towers

Ranging in height from 25 to 125 ft (7.6 to 38 m), monopoles consist of a single pole, approximately 3 ft (1 m) in diameter at the base, narrowing to roughly 1.5 ft (0.5 m) at the top, and may support any combination of whip, panel, or dish antennas (see Fig. 6.6). Monopoles are generally used in rural areas, near freeways, or in areas where buildings are not of sufficient height to meet line-of-sight transmission requirements. In the wireless system, monopoles are used much more commonly than lattice towers.

The monopole is a versatile tubular pole developed for the worldwide cellular market, with special attention paid to environmental acceptance. Aesthetic design was considered a top priority during the development stages. The efficiency of the design allows it to be an economical option for a range of heights to up to 40 m (131 ft). The monopole can be used for microwave links and cellular and telecommunications applications.

Generally, there are two types of towers: guyed and self-supporting. Each type has different limitations regarding its capacity, stability, and versatility.

Guyed towers are slender, generally three- or four-sided lattice construction, and are uniform in dimension over their length. They are maintained in their vertical position by guy wires, which are at-

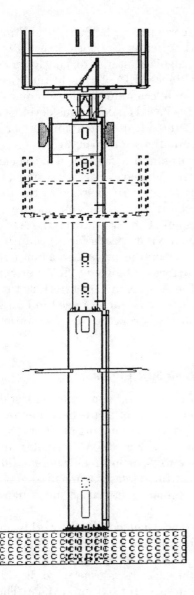

Figure 6.6 Monopole, 15 m.

tached at various levels on the tower and anchored to the concrete on the ground. Under normal circumstances, a guyed tower requires a radius of 80 percent of the height of the tower. Table 6.2 illustrates the fact that they require a large piece of land to guy a single tower, so they are usually used in rural areas. Guy stays stretch over time, and the tensions within these assemblies should be adjusted from time to time. The scope of recommended tower inspection and maintenance is published in both ANSI and CSA publications.

TABLE 6.2   Approximate Area Required for Guyed Towers

Guyed tower height (feet)	Required area (feet)
60	94 × 94
100	149 × 149
200	288 × 288
300	428 × 428
400	565 × 565

Self-supporting towers stand on their own and require a base spread of approximately 13 percent of their height. Table 6.3 shows the minimum radius of the circular surface required for the self-support tower erection. These towers are relatively inexpensive up to about 100 ft (35 m) or so but become very costly as their height increases. They are used mostly in urban areas where minimal land is available. Short self-support towers are also sometimes placed on the rooftops of the buildings.

TABLE 6.3   Approximate Area Required for Self-Supported Towers

Self-supported tower height (feet)	Required area for three-leg tower (radius in feet)	Required area for four-leg tower (radius in feet)
50	13.8	15.4
100	17.3	20.5
150	21.9	25.4
200	26.4	30.6
300	34.8	40.8

Steel (galvanized after fabrication), for the most part, is used as the material for the mast construction of towers, and galvanized guy or bridge strand is generally selected for the guy stays that support the guyed tower. The lattice mast of the tower can be made using a variety of steel shapes and connection types. A common mast makeup for a multiple microwave dish and line support tower would be angle leg, angle bracing, and bolted construction, with a climbing ladder inside the mast. This configuration offers relatively low manufacturing cost, significant strength, high stability, ease of antenna attachment, ease of maintenance, and low shipping costs. On the downside, because of the relatively high, "flat" (angle) projected area of this type of mast, such a tower and foundation system is more expensive than a comparable tower that uses round members (pipe, tube, or solid round) in its makeup. This price difference becomes smaller when the antenna and line load increase.

The most economical steel section to use for tower mast components is hollow tubes, with a circular cross section (pipe or tube) offering highest compressive capacity-to-weight ratio as well as lower wind resistance than any other shape. Canadian designers and tower manufacturers have, for the most part, excluded the use of thin-wall and small-diameter hollow sections for primary structural components (legs) in tower construction for a variety of reasons. Even properly vented sections entrap debris and water in the form of ice over time. The entrapped freezing water expands, exerting some pressure on the tube, and aids in the entrapment of debris filtering down the tube, which occasionally blocks the drain holes. Once these holes are blocked, the moisture, along with other chemicals present in it, stay trapped in the tube. The galvanized surface on the inside of the tube section is subject to potential corrosive action, depending on the acidity of the water mixture within it. An inspection of such a tower would not reveal any structural distress until a crack or hole appeared in the wall of the tube.

There are many pros and cons in material selection, connection type (welded or bolted), and mast configuration (position of transmission line brackets and ladder, facility to attach antenna mounts) for tower design. Coupled with potential differences in proposed tower accessories (grounding, safety climbing devices, anticlimb devices, waveguide bridges), towers, even given the same basic specifications, often vary considerably in price. A thorough evaluation of a proposal for the tower and the expertise and reputation of the supplier, along with the price, should be used in the selection of a tower structure.

6.9.2 Minimum Visual Impact and Other Antenna Mounting Structures

Unlike ground-wired telecommunications (e.g., the land-based telephone system), wireless communications technologies, by their operational nature, require numerous antennas to be mounted at various heights throughout the landscape. To place them at the specific height required by a particular system, these antennas are sometimes mounted on towers, monopoles, tall buildings, or other structures on hilltops. One of the greatest concerns faced by local jurisdictions is the visual impact of wireless communications facilities. A number of companies today develop and sell minimum visual impact and multifunction communication structures. Because of their aesthetically pleasing design, these alternative structures can solve unique communication requirements and enhance the permitting process in difficult zoning locations. Each structure naturally blends into the surrounding environment, and its antennas are hidden from view.

Examples of these concealed installations include structures that resemble church steeples, with crosses incorporated into the design of the monopole; lighting structures with curved arms and canisters to conceal antennas; signposts that are two- or three-legged structures with optional signage; and the ultimate design—the minimum visual impact tree. The design of the *tree pole* is site specific to the location, with careful consideration and configuration given to existing trees (see Fig. 6.7). Pine and palm tree versions are available, as well as minimum visual impact structures that resemble a saguaro cactus.

The advantage of split-configuration microwave radios is that they can be installed in a very limited space. Poles, wall mounts, and tripods are just some of the examples of antenna mounting structures that could carry the ODU and antenna.

New antenna structures on a building can take different forms and shapes. The type of antenna support structure depends on the antenna type, its height, and the existing structure. The antenna support structure may consist of a self-supported tower on a roof, a guyed tower on a roof, building walls, a tripod, or a pole.

A *nonpenetrating tripod* is often used as a temporary solution for a rapid deployment of the microwave system. For *penetrating tripods,* there usually is a requirement for architectural analysis and approval,

Figure 6.7 Monopole (in the middle) pretending to be a tree.

since drilling of the sensitive roof structure has to take place before any rooftop installation is allowed. It is important to ensure that nobody walks in front of the antenna, since that could interrupt the traffic on the microwave link and be hazardous to the person's health.

6.9.3   Maximum Allowed Antenna Deflection

It is important to remember that the requirements for the twist/sway of the tower are sometimes much more stringent for microwave than for RF installations. A typical limitation in twist/sway for the structure (tower and antenna) corresponds to a maximum 10 dB signal attenuation due to antenna misalignment. A simple formula for estimating the maximum allowed deflection (single side) for microwave transmission antennas is

$$\alpha_{-10\ dB} = \frac{60\lambda}{D} \quad \text{(degrees)}$$

or simplified even more,

$$\alpha_{-10\ dB} = \frac{18}{fD} \quad \text{(degrees)}$$

where λ = wavelength (m)
   D = antenna diameter (m)
   f = frequency (GHz)

Table 6.4 shows maximum allowed antenna deflection for some commonly used frequency bands and antenna sizes.

6.9.4   Communication Tower Requirements

6.9.4.1   Tower locations.   Three types of areas, based on their potential suitability for wireless facilities, are classified as *opportunity areas, sensitive areas,* and *avoidance areas.* It should be noted that collocation of antennae on existing towers or alternative tower structures is encouraged in all areas, including avoidance areas.

Opportunity areas are the most likely to provide good sites for the widest range of telecommunications installations, including towers. Opportunity areas include interstate highway corridors, industrial parks, shopping centers, large agricultural tracts, and other locations where properly designed facilities could fit into the landscape reasonably well and would be unlikely to become a blighting influence on the surrounding neighborhood.

TABLE 6.4   MW Antenna Deflection (–10-dB Points)

Frequency (GHz)	Antenna diameter (m)	Deflection (–10-dB points) (°)	Frequency (GHz)	Antenna diameter (m)	Deflection (–10-dB points) (°)
2	2.4	3.8	13	0.6	2.3
2	3.0	3.0	13	1.2	1.2
4	2.4	1.9	15	0.3	4.0
4	3.0	1.5	15	0.6	2.0
6	2.4	1.3	15	1.2	1.0
6	3.0	1.0	18	0.3	3.3
7	1.2	2.1	18	0.6	1.7
7	2.4	1.1	18	1.2	0.8
7	3.0	0.9	23	0.3	2.6
8	1.2	1.9	23	0.6	1.3
8	2.4	0.9	23	1.2	0.7
8	3.0	0.8	38	0.3	1.6
11	1.2	1.4	38	0.6	0.8
11	2.4	0.7			

Sensitive areas, such as high-density housing districts, sites within 500 ft of low-density residential areas, and community facilities such as churches, cemeteries, playing fields, and recreation centers, require more care in site selection, facility design, and screening. Issues such as safety, visibility, property values, and land use compatibility are more likely to arise in these areas than in opportunity areas.

Avoidance areas are the least preferred locations for wireless telecommunication towers. Low-density residential districts, ridge tops, historic sites, scenic highways, and most public parks are included in this category.

6.9.4.2 Microwave requirements. Tower site construction involves many steps: building the access road, bringing in electric and phone lines, erecting the fence and installing other security measures, providing and installing the equipment shelter, erecting the tower and installing transmission lines and antennas, installing microwave equipment, testing the microwave system, and so on. The following factors should be kept in mind during the detailed design and deployment of the MW system:

- Ensuring sufficient space on the tower to install and pan the microwave antenna

- Loading of the antenna mounting structure (MW antenna, transmission lines, and outdoor MW unit)

- Maximum allowed twist and sway (antenna deflection) of the antenna mounting structure (in degrees) (will depend on the frequency and antenna type)

Because of the complexity of existing standards, regulations, and requirements, it is very important that experienced and licensed civil engineers handle the matter of civil construction. Input for civil construction dimensioning must come after thorough transmission site surveys. The survey should produce basic requirements such as tower heights, tower stability, access road existence, and so forth. In addition, it is important to produce documentation on all levels for civil construction and installation to avoid future liability issues. The documentation should include calculations that clearly show how construction requirements are met, including maximum tolerances due to load. Site drawings for electricity, alarms, air condition, and so on are also required.

6.9.4.3 Structural requirements. At the outset, we should consider including a provision for future antennas and transmission lines in the procurement specification. At this time, it is beneficial for the owner to consider and plan for potential changes and/or additions to antenna loading that could occur in the future. The costs associated with additional tower strength would be small when compared to the high cost of reinforcing an existing tower and foundations.

Generally, if significant unplanned changes in antenna loads are made after the tower is installed, costly reinforcement will likely be required. If the changes or additions are too extensive, then the tower may have to be replaced with a larger one. Keeping that in mind, the best strategy is to create a specification for tower procurement either by using existing staff or by retaining a consultant who specializes in this service and solicit prices from several tower companies. If this approach is used, the specification for tender should include a request for sketches or drawings of the proposed tower and foundations. Towers from several different sources can vary substantially in size and, therefore, price.

The design, supply, and installation cost of a self-support tower is approximately twice the cost of an equivalent (between 15 and 76 m in height) guyed tower; beyond 76 m, the difference in cost is even greater.

In general, a guyed tower needs considerably larger ground area than a self-support tower, which might be costly in some countries.

Large ground area requires considerable fencing which also can become very costly. Guyed towers are prohibited in some countries (Germany, for example), demanding self-support towers or monopoles. However, self-support towers are generally more expensive than guyed towers, which often can be built very quickly and are much lighter because less required construction material is required. Hence, they have a smaller wind load area.

The operational dimensioning is governed by the maximum allowed antenna deflection according to the requirements calculated by transmission engineers. It is important to note that unavailability due to strong wind load does not normally occur simultaneously with flat fading. Unfortunately, exaggerated stability requirements are very common, and stability requirements stating the strongest deflection are prevalent, without any supporting calculations. The result is extremely high construction costs.

All external equipment must be properly and thoroughly mounted to avoid vibrations and resonance buildup. Resonance is generally not a problem and is easily avoided by designing a stiff construction having a resonance frequency above 10 Hz (low frequencies give the largest amplitudes).

Some important topics to consider in civil construction are quality control, construction of foundations, concrete reinforcement methods and anchor design, quality of paint and welding points, and other considerations.

Steel is extensively used for masts and towers because of its superior strength as compared to aluminum, which requires thicker constructions that result in a higher wind load. All steel constructions are to be hot-dip (zinc) galvanized for corrosion protection, at least. The effect of twisting can be accounted for by proper tower design or by using torsion protection guiding wires for supported masts.

Steel can become brittle at low temperatures. Therefore, a proper choice of metal alloy in the construction is essential to account for temperature-dependent effects. The strain in wires for supported towers creates a need for adjustment, typically after six months of operation and then regularly every few years.

Erosion around the foundation can become a problem if the site has been unwisely positioned. Local knowledge is essential for avoiding the problem of erosion. For example, a site in parts of Africa and the Middle East might seem suitable for several years but can suddenly be transformed into a river. Soil testing is also essential when building heavy constructions. It is advisable to take several soil samples for analysis. A soil test is a sample, 6 to 10 m deep, that provides information about soil composition, profile, and density. The soil test gives an indication about how to avoid long-term misalignment of the construction due to earth sliding or compression.

6.9.4.4 Relevant standards. There are a large variety of country-specific regulations and standards for civil construction. Country-specific standards for civil construction must be followed, with a suitable choice of additional standards to address customer-specific requirements. Local construction standards govern the requirements on towers and masts. These standards are often locally adapted standards with origins in British or French construction standards. Proper quality control of manufactured material and design are key factors for obtaining a high construction standard and longevity.

Minimum recommended tower strength requirements are published in structural standards or are specified by the customer. In Canada, tower strength and design are not regulated, so Canadian Standard S37, "Antenna-Supporting Structures—A National Standard of Canada," is not a legal requirement. In the United States, the tower design and construction standard is RS-222, "Structural Standards for Steel Antenna Towers and Antenna Supporting Structures—An American National Standard." A major difference between the American and Canadian standards is that the Canadian standard specifies mandatory ice and wind loads, but the American standard does not.

Some manufacturers produce a commercial, over-the-counter type of tower that is designed to criteria other than these two national standards. Such a product can provide an owner with a low-cost product for a specific and usually limited application. The incidence of failure tends to run higher with structures that do not conform to either the CSA or ANSI standards, and it is advisable to consider the risk and consequences of failure when selecting this kind of product. Usually, these types of towers have been designed to survive a specific and uniform wind velocity with no ice loading, and they have little or no safety factor beyond that loading.

The detail questions and answers for project work are to be produced and evaluated by experienced civil construction engineers in close cooperation with transmission and microwave engineers, whose requirements must be addressed.

6.9.4.5 Tower erection time lines. The typical time for erecting a mast with prefabricated concrete weights or earth anchors is less than five days. Maximum height difference between each leg is ±1 m (±3 ft). The entire mast can be erected with all radio equipment installed at ground level in one or two long sections by using a mobile crane or a helicopter. Alternatively, a hoisting device can be installed at the top of the mast for lifting additional sections and radio equipment manually.

The general time lines below can typically apply for towers under 40 m:

Digging the tower foundation	1 week
Mold and concrete reinforcement	2 weeks
Concrete filling and hardening	1 week
Pit backfilling and earth packing with watering	1 week
Tower assembly and erection	1 week
Total time	6 weeks

Laying a stable foundation is the first concern in erecting a sturdy communication tower. Once concrete is poured, a test should be performed every week to determine the foundation strength of the curing material. At seven days, the concrete should measure 70 percent of its estimated strength. Concrete reaches its normal strength after four to five weeks, depending on weather conditions. However, tower assembly on ground can start before the foundation is completed, and minor loading of the concrete is possible after one week.

6.9.4.6 Additional tower requirements. Unauthorized access to towers/masts can be prevented by fencing with barbed wire, removing ladders, installing flat plates around the lower parts of the construction, and by using different types of climbing locks.

Many types of *fall protection* can be incorporated, such as guiding tracks with wire and wheel or simply a climbing cage around the ladder. Service platforms and rest platforms are compulsory accessories in many developed countries.

The tower/mast is painted mainly for corrosion protection. A compulsory aviation warning coloring scheme is deployed in some countries. Aesthetic requirements must be considered in other countries. This can become very costly—an additional 50 to 100 percent of invested steel cost if high-quality paint is used.

Different types of aviation warning light are required in various countries, including dusk-activated relays. It is widely accepted that lightning protection can be connected to earth through the tower/mast construction.

A maximum resistance of 10 Ω to ground is dictated by British standards and is also widely observed elsewhere. However, in some countries (including the U.S.A.), more stringent criteria are sometimes applied.

6.9.4.7 Tower procurement. Under a competitive bid situation, it is the designer's task to make each tower as small as possible given the antenna and transmission line loading, code, deflection, and other customer requirements. The smallest tower, complete with its associated foundations, usually results in the lowest tendered price. The buyer of tower structures should therefore work closely with the radio design and network system planning groups (RF and transmission) in the tower tendering phase. A reputable tower design/manufacturing company is a good source of reliable information regarding pricing, scheduling, and engineering. Such businesses have engineering and technical staffs that interface closely with purchasing, manufacturing, and erection teams. Consulting firms that specialize in towers are another source of unbiased, objective information on creating a specification for towers and tower design loading. These firms are also a source of other professional services such as design, analysis, and inspection, and they can also check designs to ensure specification conformance if there exists any doubt about the proposed or installed tower.

The most vital information that has to be provided to the tower manufacturer is as follows:

1. *Tower load.* For the microwave application, tower loading of the antenna mounting structures includes antennas (immediate and future requirements), wind, ice, waveguides, coax cables, platforms, the waveguide bridge, outdoor radio units (ODUs), and so on. The detailed loading requirements should be given to the tower designer and, if climatic conditions are unknown, assumptions must be made based on local statistics.

2. *Wind speed.* Sometimes referred to as *wind load,* this is the force the wind has on the tower and antennas. The governing principles for tower/mast constructions are

   – Survival: dimension for maximum probable wind load. In Sweden, it is typically 50 to 55 m/s. The U.S. standard 50 m/s. There is also a high-wind option for up to 70 to 75 m/s.

   – Operation: dimension according to availability objectives, taking normal wind load and gusts into account. Calculation methods are comparable to microwave availability principles. Requirements are, for example, 99.995 percent availability.

3. *Ice load.*   Also known as *radial ice,* this is the amount of ice in inches formed around each tower member.

4. *Soil report.*   This report details the soil conditions present at the site and helps to determine what type of foundation is required.

5. *Other design specifications.*   These include FCC, FAA, and other specifications.

Generally, the minimum recommended requirements for tower strength are either published in structural standards or specified by the customer. These standards are constantly under review by committee and are revised and reissued from time to time to reflect the current knowledge of loading, analysis technique, materials, and workmanship.

6.10 Power Supply and Battery Backup

6.10.1 DC Power

Transmission equipment is usually required to operate off the −48 and/or +24 V power supply. In wireless networks, battery backup shall provide sufficient reserve capacity to allow the uninterrupted operation of equipment for at least four hours for the cell sites and at least eight hours for the switch office. Customers may specify more or less capacity, as local conditions such as travel time (MTTR), standby generation of power, and power reliability may vary with location. The equipment shall be capable of successfully operating within published specifications and satisfy all required operational specifications with power variations that do not exceed −56 VDC or fall below −40 VDC at any given time. In newer and more advanced power systems, alarms will notify the user when the voltages get to within 20 percent of these extreme values. In the case of battery drain, it will signal when the equipment is within one hour of losing backup power.

Both the constant-voltage and the constant-current charge methods are suitable for charging recombinant lead-acid batteries. Of the two, the most common method used is simple constant voltage, which allows the battery to seek its own current level in overcharge (or float charge). For cyclic applications in which batteries undergo regular charge and discharge, the charge voltage per cell can range from 2.4 to 2.5 V. Although a float charge of 2.25 to 2.35 V is advisable for batteries in standby use, it is important to consult the battery manufacturer's literature for proper charge-voltage recommendations. The general design criterion used to calculate necessary capacity for the battery and the rectifiers is

$$\text{Rectifier rating (A)} = \left(\frac{\text{Power consumption}}{\text{System voltage}} \right) +$$

$$\left[0.1 \times \left(\frac{\text{Installed power}}{\text{System voltage}} \right) \times \text{Charging time} \times 1.2 \right]$$

This formula shows that the rectifier has to be able to recharge empty batteries in "x hours" (usually four hours) while carrying the full load of the system indefinitely. Ideally, the system should be easy to upgrade and expand (modular approach) to accommodate changes and growth.

Example So, let us assume that the power consumption of the small microwave terminal is 1000 W. A rectifier will have to provide at least 21 A of current (1000/ 48) for correct operation of the equipment. If the battery was completely discharged, the rectifier will have to provide charging current of 10 A as well (0.1 × 21 × 4 × 1.2). The total current that the rectifier must provide at any given moment is at least 31 A.

6.10.2 Batteries

Power loss in a communication system can be disastrous for an operator's business and for an organization relying heavily on such service. The battery is by far the most unreliable part of any system that depends on it. This is true because, if any one of many external variables is incorrect, the natural deterioration process within each cell is accelerated. When the utility power fails and the battery instantly assumes the critical load, the choices and investments that have been made in emergency power equipment will be realized. The *accelerating factors* are *excessive cell temperature, too high or too low cell float voltage,* and *too many discharge cycles.*

The first step to minimize the problem is to become aware of and then track the critical variables with an automated continuous monitoring system. The second step is to correct and optimize these critical variables wherever possible. The third step is to immediately notify the user if the critical measured parameters are outside of defined limits. The final step is to automatically identify unfavorable trends and notify the users of a developing problem before it becomes serious.

Historically, backup power for telephone networks has been provided using "flooded" batteries. Over the past several years, there has been a trend toward other types of batteries such as valve-regulated lead-acid (VRLA) batteries (also called *gel cell* batteries). The term *valve-regulated* identifies a battery that is equipped with mechanical safety vents that can open under excessive overcharge. Using VRLAs (they require less attention) offers remote monitoring and reduces the number of sites that must be visited for troubleshooting and routine maintenance. This, in turn, reduces the cost of these activities.

Flooded batteries are generally too large to fit into the limited space available for wireless sites applications. Service and maintenance of flooded batteries in remote sites would also be very expensive, since the service technicians would have to visit many sites on a regular basis to check electrolyte levels. VRLA batteries do not give off any gases under normal conditions, so special ventilation provisions are not needed. Relative to flooded batteries, VRLA batteries also offer excellent power density, do not require any regular electrolyte maintenance, and can be mounted in any orientation. These characteristics make VRLA batteries ideal for these applications.

Equipping power plants with the remote monitoring and control system is a very smart investment, as it reduces both the costs and frequency of personnel visits to the cell sites. They usually generate alarm messages but also can provide vital data about power plant conditions.

6.10.3 AC Power

Although microwave equipment requires the usual Telco power (–48 VDC), there is a chance that some additional external equipment may require AC power. There are three distinct AC *uninterruptible power supply (UPS)* configurations: standby, line-interactive, and online.

A *standby UPS* (also known as an *off-line UP*") consists of a basic battery power-conversion circuit and a switch that senses irregularities in the electric utility. The load is connected directly to the utility power, and power protection is available only when there is an outage. However, some standby UPSs include suppression circuits or power line conditioners to increase the level of protection they offer.

A *line-interactive UPS* offers a higher level of performance by adding voltage regulation features to conventional standby designs. Like standby models, line-interactive UPSs protect against power surges by passing the surge voltage to the load until it hits a predetermined limit. At this point, the unit switches to the battery, but only after passing most surges through to the load. Line-interactive units can provide moderate protection against high-voltage spikes and high-frequency transients, but they do not provide complete isolation for the load.

An on-line AC UPS uses double conversion (AC/DC and DC/AC), providing complete isolation from most types of power problems.

6.10.4 Solar Energy

Photovoltaic (PV) power is a semiconductor-based technology (similar to the microchip) that involves converting light energy directly into an electric current that can either be used immediately or stored (e.g., in a battery) for later use. Photovoltaic solar cells produce electricity directly from sunlight, and they are usually made of silicon. The cells are wafer-thin circles or rectangles, about three to four inches across. Solar cells operate according to what is called the *photovoltaic effect* in which sunlight hits the surface of semiconductor material, such as silicon, and liberates electrons from the material's atoms. Certain chemicals added to the material's composition help establish a path for the freed electrons, and this creates an electrical current. Through the photovoltaic effect, a typical four-inch silicon solar cell produces about 1 W of direct current.

Photovoltaic panels/modules are very versatile and can be mounted in a variety of sizes and applications; e.g., on the roof or awning of a building, on roadside emergency phones, and as very large arrays consisting of multiple panels/modules.

Many remote sites use photovoltaic cells to generate power for onshore and offshore traffic control systems, microwave radio stations, and so on. They also provide electricity to remote cabins, villages, medical centers, and other isolated sites where the cost of photovoltaic power is less than the expense of extending cables from utility power grids or producing diesel-generated electricity.

In remote sites solar energy can be used to replace the electrical utility-based power supply entirely or just as an additional power that will help during the peak hours and reduce the electrical power consumption. In either case, for every project, a very extensive study must be done to determine the feasibility and cost efficiency of such a system.

6.11 Grounding, Lightning, and Surge Protection

6.11.1 Grounding

Grounding can be described as the science of obtaining a low-resistance path for the dissipation of current into the earth. There are different methods for obtaining a ground but, first, a discussion of grounding fundamentals is crucial to understanding and designing a grounding system. Grounding is the physical bonding or connection of equipment by a conductor to earth. Without a proper low-resistance ground, standard protection devices such as breakers and transient voltage surge and lightning protection systems are ineffective. Most communication equipment manufacturers may void their equipment warranties at sites where the ground system performance does not meet their explicit earth-grounding requirements—typically 5 Ω or less. Additionally, good grounding has other benefits, such as en-hanced personnel safety; reduction of system noise; and protection from lightning, unwanted voltages and currents, and power surges.

The earth is composed of many materials that are variously good and poor conductors of electricity, but earth as a whole is considered to be a good conductor. For this reason, and as a reference point, the earth's potential is assumed to be zero. When an object is "grounded," it, too, is thereby forced to assume the same zero potential. If the potential of the grounded object is higher or lower, current will pass through the grounding connection until the potential of the object and earth are the same. The *earth electrode* is the connection path from the equipment to the earth. The resistance of the electrode, measured in ohms, determines how quickly, and at what potential, energy is equalized. Hence, grounding is necessary to keep an object's potential the same as that of the earth's.

Grounding is critical for the human safety, and it is meant to protect equipment from voltage surges and transients. A low-resistance ground will keep equipment at or near earth potential, reducing any voltage difference between equipment and "earth." This can prevent an accident or fatality during human contact. Equipment damage (e.g., to sensitive telecommunications equipment) from surges caused by lightning and other sources can result in the loss of millions of dollars in damage and downtime.

For raw land sites, measuring soil resistivity is the first step in the planning process. With this data, a soil resistivity profile is built. This profile is the key to an accurate ground grid design and is the only way to ensure predictable grounding system performance. A soils profile is the collection of soil resistivity levels at 5-, 10-, 20-, and 40-ft depths, and sometimes up to 100 ft. This data is gathered at three or four different locations on the site. This allows the design engineer to determine the length, configuration, and quantity of rods required to achieve the specified ground-system resistance.

Shortcutting the planning stage causes many grounding installations to be found lacking on project completion and is responsible for many acceptance test failures. Equipment manufacturers and some regulatory agencies take a serious look at the installed ground-grid resistance measurements before operational approvals are granted. Soil resistivity testing allows a proper design that will save time, money, and effort.

At collocated and existing sites, the existing grounding system must be tested and evaluated. This data will be used to determine if a grounding system upgrade is required to protect the new equipment. A lack of planning and poor or nonexistent soil resistivity testing will result in unpredictable ground system performance and ineffective equipment protection. Historically, the site and ground grid testing step has been the most often undervalued one.

Some of the most commonly used grounding systems include driven rods, water pipes, chemical wells, ufer grounds,[*] and electrolytic rods.

6.11.2 Surge Suppressors

We use gas-tube surge protectors for microwaves because of their low insertion loss, low voltage standing wave ratio (VSWR), easy installation and wideband behavior. Split-configuration digital microwave radios are normally housed in weatherproof boxes intended for mounting on towers, frames, or monopoles, whereas coaxial cables are used for connecting the outdoor radios with indoor equipment. To protect sensitive microwave electronic equipment from the harmful effects of lightning transients (electromagnetic pulses) and induced voltages, in-line microwave surge arrestors (or suppressors) are used. They are usually installed in the equipment room, in the master ground bar, but they could also be installed outside the equipment room and connected directly to the external ground bar.

Coaxes carrying signals between outdoor and indoor radio units operate not only in baseband, including DC for powering the outdoor unit, but also in intermediate frequencies such as 70 or 140 MHz and even higher values. This broadband behavior prohibits the use of quarter-wave stub protectors for microwave applications, since they cannot pass DC signals, and their frequency bands are generally incompatible with those bands traveling across the IDU to the ODU coaxial. Two fundamental electrical characteristics of microwave surge arrestors include their ability to pass a DC electric current and a broadband frequency range.

The first characteristic allows the arrestor to pass the power used to supply outdoor microwave radio units, such as transceivers, high-power amplifiers, and voltage-operated switches. RF systems using tower-top amplifiers also need DC power. The second characteristic, a bandwidth of hundreds of megahertz, allows the arrestors to pass RF signals between indoor and outdoor radio units. Typical bandwidths include DC to 1 GHz, DC to 2 GHz, and DC to 2.5 GHz.

It is important to inspect all surge arrestors periodically to look for visible indications of deterioration that might suggest the need for replacement. It is also possible to develop a specific method for changing gas capsules based on statistics regarding the probability of a lightning strike to towers or monopoles of various heights in various cities.

---

[*]A ufer ground utilizes the rebar in a concrete foundation as an earth electrode.

6.12 Microwave Testing and Troubleshooting

6.12.1 Factory Acceptance Testing

Quality control is the establishment of quality standards for all materials, equipment, and services necessary for the successful execution of the project, followed by systematic measurement of quality actually achieved, comparison with the standards, and corrective action where appropriate. Quality standards should be established jointly and mutually agreed upon by the client and the equipment/service provider.

The *first article sample* is a production component or components submitted as being representative of a specific process using production tooling, equipment, methods, technique, standards, personnel, and controls.

A *first article test (FAT)* is conducted to verify that the first article samples meet the performance requirements during and after all specified environmental and durability/endurance conditions. *First article inspection* is a verification that the first article samples manufactured using the normal production process (i.e., planning, technical/work instructions, material processing systems and controls, tools, fixtures, test equipment, and personnel proficiency) will produce a component that is in compliance with all requirements. Examples of such requirements are dimensional characteristics, material content, process, capability, and performance.

Test facilities to be utilized in performance of the first article test, if known at the time of the RFQ response, shall be identified in the quotation, giving the facility name, location, contact, and a phone number. In addition, *first article test costs* are individual costs for the entire first article test program, including hardware, fixtures, equipment, test procedure development, itemized cost by test parameter, and test report preparation. These costs shall be quoted separately from the cost of production hardware and may be a part of RFQ process as well. *Test plans* are written by the supplier and/or independent test laboratory and may be amended by the customer.

During the entire duration of the microwave project, customers have the right to witness *acceptance testing* on their equipment during the manufacturing process. This is very useful during lengthy projects where there is a possibility that the quality standards could be changed over the associated period of time. Customers will definitely want to witness acceptance testing in a case in which the first shipment of the equipment possessed some problems that could be directly traced to the manufacturing process and QA. In most cases, when dealing with the reputable equipment supplier, factory acceptance testing will be limited to only the first article inspection.

If problems arise in the field, the customer can request that the supplier include additional tests that may better define the problem and point out a potential solution.

6.12.2 Field Acceptance Testing

The objective of a *field acceptance testing* document (also called the *acceptance test procedure, ATP*) is to outline the requirements and standards that client expects from its subcontractor(s) with regard to field acceptance testing of the provided microwave equipment. Such testing will ensure that all equipment is in proper working order and installed to meet the microwave link engineering specifications. This record of the standard start-up tests, defined by the *client*, agreed upon by microwave supplier, and performed after the MW radio system installation is complete, will be provided by the supplier's technician and witnessed by operator's representative (or its designated representative—subcontractor, consultant, or other individual). The system verification document should have authorizations and signatures of people involved in the project and their supervisors. The *system verification document* typically contains the following information:

1. Site-specific information and contact names

2. MW link engineering details

3. Visual mechanical and physical inspection of the MW radio equipment, cables and waveguides, antenna mounting structures, antennas, labeling, and miscellaneous equipment

4. Electrical measurements on the MW radio, typically including the following information (test results):

    – Grounding measurements

    – Antenna/waveguide/coax return loss measurements

    – Input DC voltage

    – Transmitter power verification (Output transmit power shall be measured at the waveguide output/antenna port via a microwave power meter. This value shall be the same as the designed value used in the link engineering sheet, −0.5/+2 dB.)

    – Frequency accuracy measurements—measure and record Tx/Rx local oscillator frequency + 10 ppm

    – Receiver tests including AGC characteristics (The receive signal level shall be measured at the monitor port on the indoor unit and shall equal the calculated value on the link engineering sheet ±2 dB.)

– Data service channel and VF orderwire testing

– System alarm and control

– Loopback capabilities

– System gain [Once the output transmit power has been measured, this value, along with the receiver threshold (i.e., sensitivity) of the radio on the opposite end of the link (obtained from the factory acceptance test), should be used to calculate the system gain in decibels.]

5. Check of pressurization (if installed)

6. Short-term BER measurements (1 min each port)

7. Protection switching operation test (Both ends of the microwave link must experience a manual simulated failure to verify switching to standby transmitter.)

8. Power supply redundancy verification

9. Modules and inter-rack cable continuity

10. Coax, ABAM, and FO cables and cross-connects' continuity

11. User interface functionality

12. Spare module testing

13. Battery backup time for the microwave system

14. Network management system and craft interface verification

15. Long-term (24-hr) BER testing (per hop)

16. Long-term (24-hr) BER testing (for the system, if applicable)

The document should also contain information that will allow technicians to start troubleshooting the problems and receive immediate technical support, as follows:

1. Emergency technical support hotline phone numbers. Examples of emergency technical support include service-affecting problems reported by either side and/or hardware failures that cause an outage or degradation.

2. User manuals and as-built documentation.

3. Warranty, repair, and return procedure description.

A record of the standard turn-up tests has to be provided for a number of reasons. The first and the most obvious one is that it is a part of the overall transmission network acceptance procedure and a proof that the system is functional. The other reason is that it will be the benchmark for all future troubleshooting of the transmission network. During maintenance tests on microwave systems, the results will always be compared against the acceptance testing results. Any deviation from in these results should be noted or further investigated.

It is also a good practice to include a copy of the manufacturer's product information that lists the test procedure along with the actual test results and a sketch of how the tests were conducted.

In performing tests while bringing a system into service according to Recommendation ITU-R F.1330, it is desirable to avoid the times of year and times of day when multipath propagation is most likely to occur. Studies carried out in temperate climates indicate that multipath propagation effects are least likely to occur in winter and in the two preceding months. For tests that must be carried out in summer, the period during the day when such effects were observed to be least likely was 1000 to 1400 hours, local time. It is reasonable to assume that the same is true in other seasons.

6.12.3 Bit Error Rate Testing (BERT)

The total time in a digital transmission system is divided into two categories: *available* and *unavailable* time.[8] A system becomes unavailable if the BER is equal to or worse than 10^{-3} for more than ten consecutive seconds. The following error performance parameters are used in describing transmission quality:

- Unavailable seconds

- Percent availability

- Severely errored seconds

- Percent severely errored seconds

Live data emulation is a pseudorandom pattern that uses a quasi-random signal sequence (QRSS). This pattern provides a good approximation of live traffic with an approximately 50 percent ones density. This pattern generates all possible combinations of a 20-bit binary counter except *All Zero* and is limited to a maximum of 14 zeros.

QRSS tests are very useful to verify connectivity, but the system also has to be tested with a set of tests to make sure that there are no hidden issues (e.g., AMI-B8ZS mismatch, faulty equipment) that could cause intermittent problems later, during operation. This is specially important in mixed microwave and leased line environments, and end-to-end tests should also include *All Ones* (a fixed test pattern of pulses only), *One in One* (1:1, alternating ones and zeros), *All Zeros* (used to test circuits for clear channel capability), and stress testing. The most commonly used fixed pattern is 3 in 24. The 3 in 24 pattern simultaneously stresses the minimum ones density and the minimum number of consecutive zeros criteria.

After installation of a microwave system, out-of-service BER testing is the most useful tool for verifying equipment operation and end-to-end transmission quality. Out-of-service testing allows the user to accurately assess the quality of a T1/E1 and/or T3/E3 circuit by transmitting and analyzing test patterns in place of the live data that is normally present. Because all out-of-service testing requires that live, revenue-generating traffic be interrupted, it is impractical for long-term testing. Thus, this type of testing is typically performed when a system is initially installed or when errors are discovered while monitoring data.

Two methods of out-of-service testing are typically used to analyze T1/E1 networks: *end-to-end testing* (two test sets required) and *loopback testing* (one test set required). End-to-end testing is performed with two test sets, so the analysis can be done simultaneously in both directions, and the direction of errors can be found much faster. The recommendation is to avoid loopback testing and perform end-to-end testing whenever possible. By simultaneously generating a test data pattern and analyzing the received data for errors, the test instruments can analyze the performance of the link in both directions.

For normal transmission facilities, acceptance tests are commonly run for periods of a few hours at most. However, for microwave systems, it is a good idea to perform *long-term tests*. Tests should be performed over a period of a few days to cover different atmospheric and weather conditions and ensure that the microwave system will operate well under all these conditions.

The *in-service method* allows live data to be monitored at various access points without disturbing revenue-generating traffic. Because in-service monitoring does not disrupt the transmission of live traffic, it is more suitable for routine maintenance than out-of-service testing.

6.12.4   Troubleshooting Microwave Systems

As with any other system, problems on a microwave link can be intermittent or repetitive in nature. They can also be caused by propagation problems, incorrect design and installation of the equipment, and faulty modules within the radio itself. Long-term test results from

BERT and/or network management system will narrow down potential causes of the microwave link performance degradation.

Printouts of the long-term dynamic performance test report have to have a precise date and time stamp to correlated the problem with other events—weather fronts and rain storms, maintenance and manual intervention, software upgrades, switching, fading, local airport flight times, power issues, and so forth.

6.13 References

1. Federal Communications Commission, *Evaluating Compliance with FCC Guidelines for Human Exposure to RF Electromagnetic Fields,* OET Bulletin, Edition 97-01, August 1997.
2. Safety Code 6, Health Canada, "Limits of Human Exposure to Radiofrequency Electromagnetic Fields in the Frequency Range from 3 kHz to 300 GHz" (also available from http://www.hc-sc-gc.ca).
3. Federal Communications Commission/Office of Engineering and Technology, "Evaluating Compliance with FCC Guidelines for Human Exposure to Radiofrequency Electromagnetic Fields," OET Bulletin 65, August 1997.
4. Institute of Electrical and Electronic Engineers, "IEEE Recommended Practice for the Measurement of Potentially Hazardous Electromagnetic Fields—RF and Microwave," IEEE STD C95.3-1991, 1991.
5. Andrew Corporation, Catalog 38.
6. Gang Liu et al., "The Effects of Wet Radome on a Short Millimetre-Wave Link in Singapore," School of Electrical and Electronic Engineering, Nanyang Technological University, Singapore, 2000.
7. Kaplan, E. D., Ed., *Understanding GPS: Principles and Applications,* Norwood, MA: Artech House, 1996.
8. Gruber, J., and Williams G., "Transmission Performance of Evolving Telecommunications Networks," Norwood, MA: Artech House, 1992.

Project Management

7.1 Tracking Microwave Rollout

7.1.1 Project Management Activities

Project management is the application of knowledge, skills, tools, and techniques to project activities so as to meet (or exceed) customer needs and expectations from a project. Meeting or exceeding customer needs and expectations invariably involves balancing competing demands, scope, time, cost, and quality.[1]

A *project* is defined as an undertaking of a nonroutine (unique), nonrepetitive nature having prescribed objectives in terms of scope, time, quality, and cost. Within the realm of project management, such projects can be further defined as generally being complex, having a multidisciplinary involvement, and having various phases in their life span. The various phases of a microwave (and any other) network build-out project may be defined differently by different organizations, but they generally fall into the following categories:

- Concept (or feasibility) analysis
- Network planning and preliminary design
- Detailed design and engineering
- Deployment (also called *implementation*)
- Testing and commissioning

Each of these phases can be divided into a whole list of subphases and activities. The completion of each of these project phases is usually accompanied by a finished, smaller project of some sort. While project management skills are quite distinct from engineering design skills, the requirements of good project management are not different from the requirements of good engineering or good management.

Effective management of a project calls for the early establishment of *policies and procedures* for its implementation. During the initial phase of the project, therefore, the project manager, in conjunction with the client and other interested parties, would establish clearly defined and properly documented project policies and procedures, as well as setting design and deployment standards that meet the client's operational requirements and satisfy the needs of effective management and accountability.

Project management services are applicable in all phases of a project, from the initial concept through implementation, to the final commissioning and handover of an operational project.[2] It is therefore important that a client be aware of the full scope of services that can be provided. Project management services would normally include certain *basic activities* such as the following:

- Planning and scheduling (time management)
- Budgeting and estimating
- Cost control and accounting
- Quality control
- Regular reporting

In addition to the basic services, depending on the particular project, project management services could also include a number of other items such as:

- Interpreting the client's requirements, operational needs, and constraints and advising the client as to the suitability of alternative solutions
- Defining the project requirements, including scope, quality, and overall budget and schedule of work
- Preparing project policies and procedures
- Assisting in securing project financing and arranging appropriate financial arrangements
- Advising the client of required decisions in relation to legal and insurance considerations
- Advising and assisting the client with respect to the regulatory and approval process with statutory authorities and obtaining the required permits
- Structuring the project into manageable sub-entities

- Prequalifying, recommending, selecting, and negotiating contracts with consultants, vendors, and contractors

- Managing the design for conformity with the agreed project requirements and budget, and administering design changes

- Suggesting alternatives, evaluating them, and assisting the client in choosing among them so as to best meet the needs of the client in terms of scope, time, quality, and cost

- Identifying to the client the impact (scope, time, quality, cost) of proposed changes so that the client may make well informed decisions about whether to proceed with the proposed changes, and advising the client of the effect on the project of delayed decisions/approvals

- Arranging and coordinating the procurement, expediting, and quality control of all required materials, equipment, and services, including those supplied by the client

- Procuring equipment and services, including prequalification, tendering, contract negotiation, contract administration, and expediting

- Managing construction/implementation/deployment for conformity with the approved design, including detailed scheduling and coordination, management of inspection, administration of construction changes, approvals of progress claims, completion certificates, management of deficiency and warranty work, commissions, operating manuals, and record documentation

- Assisting the client in testing and commissioning, start-up, and/or operating procedures, including staff training

In defining the scope of services to be provided, the client and the project manager should review the above listing in detail, consider additional items that might be appropriate in the particular case, and establish the items, and the related scope, to be included in the services contract between them.

The project policies and procedures should be specifically developed to suit the size, complexity, and scope of the particular project and would normally cover overall project implementation policies, project organization, personnel functions, administrative and control procedures, technical design criteria, documentation standards, quality control procedures, and record keeping.[3]

7.1.2 Project Kick-Off Meeting

The *project manager* is most likely the first person appointed if the new microwave system is even remotely considered. The project manager will have to understand basics of the microwave system build-out process, know which specialists need to be brought into the project, and understand basic time line and critical path activities. Regardless of whether it is a new microwave system or an upgrade or expansion of the existing facilities, microwave deployment (build-out) is a multi-disciplinary activity that involves a number of very specialized experts in their respective fields.

It is always a good idea to organize a kick-off meeting and invite all the parties involved in the project. This is a good opportunity for the people to meet and see who else will be working on the project. In addition, this is a good time to identify some "missing links" in the project, i.e., equipment, experts, or anything else might have been omitted by the project manager or previously thought not to be required. This is a "brainstorming" session organized by the project manager, but it requires active involvement by all team members.

The *project system engineer* should be present at this meeting, since this person will be responsible for ensuring that the design and installation is performed according to industry, customer, and supplier standards and practices. The project system engineer is the key technical person on the project and has an overall responsibility for the technical integrity of the system design provided.

7.1.3 Network Planning

Network planning (also called *preliminary design*) is usually done in a matter of days and typically consists of the following activities:

- Discussion of the client's requirements and project goals. The client defines the system requirements and, often, site locations as well.

- Preliminary path engineering using topographical maps and visiting only key sites (switch office, big hub sites, and so forth).

- Identification of material requirements and creation of preliminary BOM.

- Development of the budget and schedule.

Preliminary path engineering, including routing design and preliminary path analysis, should be done prior to any field trips and detailed design. Computer programs are available to plot an initial system map that shows the sites in geographical relationship to each other. Approximate site coordinates and the feasibility of various paths are determined from a topographical maps, from a digital terrain database (minimum 1:50,000), or from old survey information. This map is used, along with traffic requirements, to define the paths between sites and the type of radio for each hop.

Results of the preliminary network planning activities and its deliverables (budget, work force, and schedule) form the basis on which the whole project will be assessed and potentially approved or rejected. Of course, due to the short time available as well as very limited input data, the margin of error in these calculations and predictions is quite high.

7.1.4 Project Approval

For an engineer, this is probably the most difficult part of any project. It includes convincing nontechnical personnel/executives who may or may not have the budget, may or may not have any experience with the technology involved, and probably are too afraid to commit to spending millions of dollars for something that won't start bringing any profit for at least two or three years. This is the point at which people with vision stand out from those who are concerned only with the next quarterly results.

Usually, transmission (transport) facilities are either leased or owned (copper, microwave, fiber optic) or (most likely) a combination of leased and owned (usually microwave) facilities. In most rural areas and third-world countries, there are very few options but to build the microwave system. In North America, however, leased lines are widely available and therefore a good option to consider. The only problem is that this option, although cheap if we consider only the monthly rate, becomes very expensive after few years, because it is a recurring cost that continues *ad infinitum.*

In many cases, project managers and those involved in the time-to-market assessment conclude that the faster way to build the network is to lease T1/E1 circuits from the local telephone companies rather than building their own microwave system. That may not be the case in every situation, and it may prove, for a number of reasons, to be a much more expensive and lengthy process than originally anticipated.

Building the microwave system from the ground up involves many different activities, some of which are very lengthy and very expensive. A business case has to be prepared with a due diligence, based on the preliminary microwave network plan, and it cannot be done in a day or two. If hundreds of microwave hops are required, it may take a few weeks for an experienced engineer to put all numbers and different scenarios together.

Vendor *financing* is one of the most commonly used methods for deferring the initial blow of a huge capital expenditure on a day one. Many vendors are willing to provide financing, assuming they are interested in doing business with a certain customer in a certain part of the world.

7.1.5 Site Acquisition

Everyone wants to use wireless phones, wireless LANs, broadband Internet services, and so on, but few people want the towers and antennas in their neighborhood. With increasingly restrictive zoning requirements and a host of new wireless operators rushing to install thousands of new cell sites in every large city, tools and processes are being developed to increase flexibility in site locations without deteriorating coverage, capacity, and service quality.

Site acquisition is a main bottleneck and a critical path in most wireless network build-out projects, and it can take up to six months or longer to acquire the appropriate site. The same goes for microwave-only sites.

The real estate group has to make decision based on the potential of the site to be leased, time required to finalize all the required paperwork, zoning restrictions, and other factors. In wireless and/or microwave networks, unfortunately, the optimum site frequently is not available, so compromise is required. Lease negotiations take place after the property for the cell site is selected, and the contractor must negotiate the deal with the property owner, which can be very profitable for the property owner. In accommodating and antenna(s) and auxiliary equipment, the owner could receive a few hundred dollars per month in rural areas and up to few thousand dollars per month in urban settings.

Some of the things to keep in mind during the site acquisition process that will help to speed up the process are noted below:

- Monopoles are often aesthetic and economical alternatives to self-supporting and guyed towers, and they are more acceptable to planning and zoning committees. They may not be suitable if the loading of the tower is substantial or requirement for twist and sway is very stringent (microwave radios).

- Make all legal documents as simple as possible; long, complicated documents written in language that is not easily understood may cause the property owner to be wary. The document has to satisfy legal requirements and yet allow the nonprofessional to understand it.

- Address all local community concerns regarding aesthetics, safety, potential health hazards, environmental impact, and so forth as soon and as precisely as possible.

- In dealing with the public, it is a good idea to avoid using term *microwave*, because it makes many people very nervous. Terms like *directional RF antenna* and *RF parabolic antenna*, which are also technically correct, can replace the term *microwave antenna*.

Zoning issues will be different for each jurisdiction, since every city, town, and municipality has its own particular requirements. Negotiating with public officials and civic administrators requires considerable time and experience, so operators quite often hire professionals to do the site acquisition.

Many operators optimistically set aside only a few months for acquiring, permitting, and building their initial set of transmission facilities in the effort to launch a new wireless service in a community. They are disappointed when they realize that they did not perform sufficient preliminary investigation, and they end up trying to build cell sites in upscale or historical areas where it is impossible to erect a telecommunications tower.

7.1.6   Detailed Network Design

After the initial planning phase, budget approval, and personnel mobilization, all the prerequisites for the detailed microwave design have been achieved. A microwave radio system requires careful planning and analysis prior to equipment installation. A poorly designed path may result in periods of system outages, increased system latency, decreased throughput, or a complete failure to communicate across the link. Detailed design consists of these main activities:

- Conducting site and path surveys
- Performing link engineering
- Performing interference analysis and radio licensing
- Finalizing the equipment specification (bill of materials, BoM)

Site and path surveys and link engineering are mutually dependent, as not all the sites will be suitable for tower and microwave antenna installation, and any changes in site location can seriously affect the network topology and design. A field path survey should determine the exact coordinates of locations where the antennas will be installed, establish the height of each antenna, identify the location and height of current and future path obstructions (for example, a tree may grow to obstruct the path at some later date), and identify the location of possible reflection points.

Once the site survey and field path survey have been completed, the final link engineering will be performed. Results of the link engineering process are forwarded to the company that is designing towers, shelters, and other infrastructure, and this information will serve as input data for their design.

Interference analysis and frequency coordination play a very important part in proposed route design. Governments usually require users of the radio spectrum to frequency-coordinate their planned and existing microwave radio systems with other users of the radio frequency spectrum. Such coordination is a prerequisite in any microwave radio license application submitted by a microwave radio system operator. The *frequency-assignment process* varies from country to country. In some countries, the operator will be assigned a frequency band and can then plan the frequency assignment for each hop without asking the authorities for permission. In that case, the operator is also responsible for interference analysis. In other countries, the operator has to apply for frequencies on a per-hop basis. In that case, the authority is responsible for the final interference analysis.

Interference typically affects both the "interferee" and the "interferer." An established system encountering interference may be interrupted, but the new system causing the interference often will not be able to fully or properly function. *A new system cannot displace an established system.* Once a system is operational, any future systems will have to "steer around" the existing signals so that both systems can live in harmony. When one encounters interference while deploying a new system, it is frequently possible for the established system and the new systems to arrive at some sort of mutual accommodation, enabling both to coexist while avoiding interference. This can be done through tactics such as reducing the transmitting power of one link so it does not interfere with the other link.

The objective of the *equipment specification* is to produce an equipment BoM for a proposed network design, including all details of the transmission equipment required to construct the network. The activity will enable correct ordering of equipment, thus ensuring correct equipment availability at the time of installation. It is important to include installation equipment, installation tools, and other equipment necessary so as to complete the implementation. The BoM has to be adequate to enable forecasting and ordering (procurement) of the equipment.

Any necessary network management system (NMS) equipment and installation services should also be included in the scope of the work. This part is usually handled by a group of people other than those involved in the radio part of the project.

7.1.7   Equipment and Services Procurement

Procurement is the systematic execution of the procedure for purchasing all materials, equipment, and services needed for the project, in good time, and in a manner that is cost effective. These would generally include (but may not be limited to) those provided by consultants, testing services, suppliers, construction managers, and contractors. Equipment may include all or some of the following items:

- Tower/tower upgrades
- Antennas and transmission lines
- MW radio equipment
- Other transmission equipment

Services may include one or more of the following activities:

- Transmission/microwave network design
  - Network planning
  - LOS verification and/or path surveys
  - Detailed design
- Installation services
- Project management
- Testing

Services may also include post-installation operation and maintenance of the transmission network.

For large project, two or more vendors usually are selected, and each will share the market and provide backup in case the other has problems with delivering on time or providing agreed-upon quality.[4] A decision is made based on the request for quotation (RFQ) response and its compliance with the functional and technical requirements of the client.

7.1.7.1 Answering proposals. The RFQ (or *tender document*) response describes, in detail, the equipment and/or services to be supplied. The RFQ is prepared by the customer for the purpose of soliciting hardware, software, and/or services information for evaluation and possible procurement, with a specific project in mind. Answers to questions asked in the RFQ will provide the client with a better understanding, both in financial terms and in view of system integration and capacity aspects, of the equipment and services that the supplier (vendor) can provide. Topics discussed in the RFQ are usually, but are not limited to, *commercial conditions of contract, technical conditions, project management, quality assurance and reliability issues, procurement and delivery issues, training and documentation, in-service date, RFQ response due date,* and so forth.

The *request for information (RFI)* and *request for pricing (RFP)* are somewhat less detailed requests that are usually sent to equipment or service providers to solicit information on their products and services. A response to the RFI could be just a collection of data sheets, brochures, user manuals, and similar items. The response to an RFP could consist of a few pages of standard list pricing, usually without any discount or additional considerations.

The RFQ can be functional or technical in nature. A *functional RFQ* is one in which the client describes the system and its functional requirements, and it is the total responsibility of the supplier to make it work (turnkey project). For turnkey contracts, a specific scope-of-work document is also included to define the installation and test services to be covered by the contract. In some cases, different contractors may provide original equipment manufacturer (OEM) and installation services.

A *technical RFQ* is very similar except that the client will provide more initial data to the supplier. For example, perhaps a preliminary microwave plan has been done and tentative site locations identified, which will give all the bidders the same starting point. This will eliminate a big discrepancy in the initial approach among bidders and therefore provide more realistic and competitive pricing.

If the client takes full responsibility for the microwave network build-out and only needs the equipment, a *bid specification* is created and sent to suppliers to provide a quote. Installation services may or may not be a part of the bid specification. Sometimes, on large networks, suppliers will install part of the network and train clients' personnel at the same time. After the initial few hops, the supplier's staff will leave, and the clients' technicians will complete the rest of the project.

In many transmission networks, speed of deployment will be very critical factor in the process of equipment and/or supplier evaluation, and it must be addressed and discussed in detail. All suppliers are usually provided with the opportunity to individually discuss RFQ proposals; after that, final discussions will be conducted with up to three top candidates, after which the financial and legal terms will be determined.

It is very important to use the proper terminology while preparing proposals. For example, "shall" and "shall not" identify requirements to be followed strictly and from which no deviation is permitted. "Should" and "should not" indicate that one of several possibilities is recommended as particularly suitable, without mentioning or excluding others, that a certain course of action is preferred but not necessarily required, or that (in the negative form) a certain possibility or course of action is discouraged but not prohibited. "May" and "need not" indicate a course of action that is permissible within the limits of the document. "Can" and "cannot" are used for statements of possibility and capability, whether material, physical, or causal.

7.1.7.2 Proposal pricing model.

A *pricing model* is usually defined by the client, and it has to be as close to the planned network as possible. The supplier should try to adhere to the requirements and definitions provided in the RFQ as closely as possible. Compliance of the equipment with all of the applicable national and international telecommunications and quality standards, and interoperability with the equipment of other suppliers, is usually mandatory.

Equipment evaluation is usually based not only on the technical specs and price, but also on the other criteria (e.g., the experience of other customers in the same or other countries, with the same or similar type of equipment, warranty, and customer support). Long-term maintenance costs could exhibit large variations, and it is important to consider this aspect during the system comparison.

The RFQ should be structured to simplify the process for the client and supplier alike; responses must specifically address each point set forth in the RFQ and should be clearly answered in the response tables. The amount of information submitted is left to the discretion of the respondent, but it is imperative that pertinent information be submitted and individual topics of interest dealt with completely and concisely. Those suppliers failing to provide complete and accurate responses can be discredited for the quality of the response and appropriately penalized in the *response evaluation* process. A supplier who chooses a *no-bid* response to the RFQ should specify, in a cover letter, the reasons for the decision.

The usual response time, depending on the complexity of the project and client's requirements, is two weeks for very small projects and up to eight weeks for large microwave networks. The more details about the future network are provided to suppliers, and the more time they have to reply, the better are the chances that their estimates will be close to reality, with fewer surprises to follow.

The quotes are usually based on uninterrupted/contiguous site field activities. Additional mobilization costs are not included in the pricing, and if the installation is delayed due to inclement weather, an inaccessible site, or incomplete site preparation or construction by others, additional charges may apply.

7.1.8 Site Ready for Installation

In this context, *site ready for installation* means that the site is ready for the microwave equipment to be delivered, installed, tested and commissioned. One of the basic assumptions (after establishing LOS, of course) is that there is available space to install the microwave radio, antenna, and all other miscellaneous equipment, and that AC power (and maybe even a DC charger and battery backup) is readily available. Access through site security has to be facilitated so that the installation crew does not waste valuable time waiting to be let into the site.

It is important to remember that, to install, test, and commission the microwave link, both sites of the link have to be ready for installation. Although it can be done, installing equipment on one site one week and then waiting a few months to finish the installation on the other site that completes the link is not recommended. It is a waste of time and resources and, on a large project with many remote sites, it can amount to a sizeable increase in installation costs.

It must first be noted that operators' capital expenditures consist of civil engineering outlays (network construction and other services) as well as investments in telecommunications equipment. The civil engineering portion, although varying tremendously according to the type of project and location of network deployment, seems to average approximately 60 to 70 percent of the cost for the mobile system gear. While labor costs—the bulk of the civil engineering expenses—are unlikely to fall, there are strong reasons to believe that, as a proportion of total capital expenditures, they will actually decline in coming years. New technology, after all, does not only offer higher capacity and lower unit cost; increasingly, it offers more efficient operations.

7.1.9 Equipment Installation

Telecommunications equipment comes in all shapes and sizes, so shelters that house such equipment often are custom designed and custom built to properly protect it. New split-configuration microwave radios require very little space and can be installed virtually anywhere. It is important to notice that, after weeks and months of site identification (site acquisition), site construction and equipment installation typically take only a few weeks. These phases include the following:

- Tower erection/upgrade

- Installing antenna and transmission lines

- Installing radio equipment

- Installing other transmission equipment

On split-configuration microwave radios, the same team will install radio, coax cables, antennas, and even the power supply. On large backbone microwave systems, a radio supplier usually provides people to install radios, but large antennas, waveguides, pressurization equipment, and so forth are installed and tested by a different team of people who specialize in this kind of work.

Antenna mounting structures could include towers, poles, tripods, walls, and others. During the detailed design and deployment of the MW system, the following are important considerations:

- The existence of sufficient space on the tower to install and pan the microwave antenna

- Loading of the antenna mounting structure (the MW antenna, transmission lines, and the outdoor MW unit)

- Maximum allowed twist and sway (antenna deflection) of the antenna mounting structure (in degrees); depends on the frequency and antenna type

- Availability of experienced civil engineers who can handle the complexity of existing standards, regulations, and requirements involved in civil construction

It is also very important to keep in mind these safety precautions:

- All rigging is to be done using safe work practices.

- All riggers must have an approved climbing certificate.

- No tower climbing is to take place until the rigging company is satisfied that it is safe to do so.

As physical equipment becomes progressively smaller and hence easier to install and commission, huge productivity gains are achieved. Infrastructure vendors already are taking advantage of that trend by preassembling and pretesting equipment at manufacturing sites before they are shipped to the installation site. This is designed to minimize costs (and disruptions) in the field. Microwave equipment has also been greatly improved by utilizing split configurations and replacing waveguides with coax cables in many applications.

It is a good practice to assign a *material coordinator* the responsibility of coordinating shipments and deliveries for all equipment and materials on the project. Material coordination is an important key to a successful and efficient implementation plan.

7.1.10   Acceptance Testing

A record of the standard field turn-up tests (defined by the operator, agreed upon by microwave supplier, and performed after the MW radio system installation is complete) has to be provide, as well as the *system verification document*. See Chap. 6, Sec. 6.12, for more details on the field tests and deliverables.

7.1.11   As-Built Documentation

An *as-built* document is usually not completed until each site is integrated; however, it should be stressed that an agreement of its content must be included in the contract at the beginning of the project. It is advisable to prepare a standard installation drawing or a real site and present it to the customer so as to minimize potential misunderstandings later. The document should include the following items (note that this refers to microwave sites only):

- *Site situation plan*—shows the site location on a map.
- *Floor plan drawing*—indicates the location of the installed transmission equipment.
- *Cable way drawing*—provides indoor cable installation information.
- *Antenna placement information*—shows the antenna arrangement—height, orientation, polarization, and so on.
- *Alarm allocation table*—indicates alarm cabling.
- *Power distribution*—indicates power distribution for the indoor unit.
- *Transmission configuration data*—provides microwave link information used for software setup.
- *Rack layout*—shows the layout of the indoor equipment in the transmission rack.
- *Transmission traffic layout*—indicates traffic distribution on T1/E1 level (also called T1/E1 plan).
- Transmission trunking diagram.
- *Plant specification*—a list of equipment used on the site, including installation material (e.g., cables).
- *Product list*—a list of main units, including equipment serial numbers.
- *Acceptance test document*—confirms the customer's site acceptance.
- Factory test results.

The *as-built document* shows how the equipment is installed in a site. From its contents, it is clear that all installation aspects are considered at this stage.

7.1.12 Commissioning

We can turn the system on when the last two segments are completed; i.e., *as-built documentation* is ready and system acceptance tests have been performed. *Commissioning* is the process of systematically bringing the various components of the project into an operational mode prior to start-up and executing a formal handover to the client. Commissioning of high-capacity links (backbone systems), hub sites, and ring sites should be performed first. After that, lower-capacity and spur sites should be turned on and commissioned. The client's staff that will operate the completed network normally carry out commissioning and start-up.

As soon as the equipment becomes operational and tested according to the ATP, the project manager would initiate and submit to the client a project final certificate for acceptance. This certificate would normally record the formal handover of the completed project and identify all the documentation (including operational and maintenance manuals, as-built drawings, equipment warranties, and contract completion reports) that the client requires for ongoing operation.

7.1.13 Maintenance Program

Although a properly designed and installed microwave system does not require a great deal of maintenance, a periodic maintenance program has to be implemented. The program has to be established during the microwave network-planning phase. Maintenance can be performed by the operator, or it can be contracted out to a third-party company. In preparing a maintenance program, the following must be done:

- Define a maintenance system and arrange for the staffing of maintenance personnel.

- Establish maintenance centers and determine the types and number of test equipment, tools, and spare parts needed for each center.

- Prepare a maintenance schedule and forms for maintenance logs and records.

- Define the target system performance values for providing maintenance and prepare standardized test procedures.

A good NMS, purchased and installed with the microwave radio equipment, will simplify the maintenance tasks and automate the process of periodic maintenance logging and record keeping. The periodic routine maintenance for the equipment is performed in accordance with the relevant equipment manuals and manufacturer's requirements.

7.2 Regulatory Issues

7.2.1 FAA and FCC

Under authority granted by the Federal Aviation Act, the Federal Aviation Administration (FAA) has jurisdiction over the following communication facilities:

- Towers that exceed 200 ft in height
- Towers that are located within 20,000 ft of a major commercial or military airport
- Towers that are located within 10,000 ft of a general aviation airport

The FAA reviews the location and height of such towers and may require them to be painted and/or illuminated to prevent possible interference with nearby airport operations. The FAA also reviews possible interference issues with aircraft-to-ground communications that may be caused by transmission facilities located in or near airport flight paths. In the U.S.A., owners of antenna towers that are taller than 200 ft (60 m) above the ground level, or that may intersect the flight pathways of a nearby airport, must register the structure with the Federal Communications Commission (FCC) and have the structure's placement studied by the FAA. Previously, antenna registration was the duty of the antenna site licensee, but now (since 1995) the owner of the real property is accountable.

FCC Form 854, "Application for Antenna Structure Registration," is to be used to register structures used for wire or radio communication service in any area where radio services are regulated by the commission. In addition, it is used to make changes to existing registered structures or pending applications, or to notify the commission of the completion of construction or dismantlement of structures.

Tower lighting is another of the very important requirements generated by the FAA. Unreported tower light outages can cost a tower owner a lot of money in the form of fines levied by the FCC. However, reducing the potential for a serious aviation accident is enough motivation for reputable tower owners and communications system licensees to maintain tower lighting in good working condition.

The FCC is an independent federal regulatory agency that is directly responsible to Congress. Established by the Communications Act of 1934, it is charged with regulating interstate and international communications by radio, television, wire, satellite, and cable, and its jurisdiction covers the 50 states, the District of Columbia, and U.S. possessions. The general objectives of federal telecommunications regulations are to provide efficient use of the electromagnetic spectrum (which is considered to be a public resource) and to develop a domestic telecommunications infrastructure that is able to provide service on the national level and compete on the global level.

In some circumstances, tower installations must be approved by the FAA, registered with the FCC, or both. To ensure compliance, it is important to review the current FCC regulations regarding antenna structures. These regulations (along with examples) are available on the FCC web site at www.fcc.gov/wtb/antenna/.

In January 2003, the FCC and the Commerce Department's National Telecommunications and Information Administration (NTIA) executed a new *Memorandum of Understanding (MOU)* on spectrum coordination. The MOU applies to the coordination of spectrum issues involving both federal and nonfederal users. The FCC is responsible for nonfederal users (e.g. broadcast, commercial, public safety, and state and local government users), and NTIA is responsible for federal users. The majority of the spectrum is shared between federal and nonfederal users, in which case the FCC and NTIA must coordinate spectrum policy.

Other limitations imposed by authorities can also affect microwave radio deployment—e.g., tower height restrictions or limitations on antenna size. These factors can restrict effective radio lengths at the planning stage and should be ascertained in advance of the detailed link design stage.

Most *local governmental agencies* regulate wireless communications facilities via land use regulations contained in respective zoning ordinances and general plans. They are responsible for reviewing and processing applications for discretionary and ministerial permits for such facilities. Local governments also have the broad authority to ensure the public health, safety, and welfare of their citizens. *Local jurisdictions* regulate wireless communications facilities through the permitting process. Most agencies require a discretionary permit, such as a conditional use permit, to construct a facility. Whether a permit is processed administratively or requires a public hearing varies among local agencies. In general, administrative processing entails lower permit fees and shorter processing times, whereas the public hearing process involves higher permit costs and a longer permit turnaround time.

The FCC's Wireless Telecommunications Bureau (WTB) handles all FCC domestic wireless telecommunications programs and policies except those involving satellite communications. Wireless communications services include cellular telephone, paging, personal communications services, public safety, and other commercial and private radio services. The WTB regulates wireless telecommunications providers and licenses with the intent of preventing interference and conflicts among various operators and services at a given location when such operators use the same portion of the frequency spectrum.

7.2.2 FCC Frequency Coordination Process

The FCC *frequency coordination process* involves several distinct but interrelated steps: *interference analysis, notification,* and *response.* After the frequency coordination process is completed, the licensee is responsible for maintaining records for the FCC after the license is granted. The licensee must notify the FCC of any address or other administrative changes to the system. The licensee must also track when a license is due for renewal (the current renewal period is ten years) and submit a timely renewal application.

The first step in the frequency coordination process is *interference analysis*. The FCC requires that applicants engineering a new system or making modifications to an existing system must conduct appropriate studies and analyses to avoid interference to other users in excess of permissible levels. This interference analysis is performed by the applicant before issuing a prior coordination notice (PCN) and is performed by recipients of a PCN to verify noninterference.

Interference analysis is an iterative process that involves computer simulation of potential interference and an engineering analysis to eliminate interference cases. The process begins with a tentative frequency selection that is consistent with the established frequency plan. Telecommunication Industry Association (TIA) Bulletin 10 criteria and industry-developed guidelines are used for automated calculations. These calculations include co-channel and adjacent channel interference, threshold degradation, adjacent spectrum interference; and potential interference from intermodulation products. In frequency bands shared with satellite Earth stations, an interference analysis is conducted with the applicable ground and space segments. All operational fixed (OF), common carrier (CC), and broadcast auxiliary point-to-point microwave operators are required to complete and submit Form 601 to the FCC. The form must be accompanied by a supplemental document indicating that the frequency coordination process has been completed. Once the appropriate application is filed, the FCC puts the application on notice for 30 days for public comment. If there is no objection, and the application meets all FCC filing requirements, the FCC grants the application.

Once an interference analysis has been completed, and prior to system implementation, an operator is required to issue the notification to all "potentially affected parties." The industry defines an operator as potentially affected if the operator's facilities (including proposed, applied-for, or operating) fall within a defined coordination distance and operate in the same frequency band. This notice is referred to as a *prior coordination notice (PCN)* and contains the technical operating parameters and a general description of the proposed system.

The recipient of a PCN has 30 days in which to analyze the proposal and provide a *response*. Every attempt should be made by the receiving party to respond as soon as possible. In most cases, operators utilize an outside agent, commonly referred to as a *protection agent,* to administer this function. The response to a PCN should include an affirmation of the proposal or, if there are objections, a detailed description of these objections. Typically, a response that raises concerns will contain enough technical data to substantiate the objection.

The party issuing the PCN is then required to resolve all potential conflicts to the satisfaction of the objecting party. This may require several rounds of discussion, technical analysis, and negotiation. When both parties have reached an agreeable resolution of the issues, the coordinator of the proposed system issues a document called a *supplemental showing*. The supplemental showing is akin to a signed affidavit in which the coordinator attests to satisfactorily completing coordination.

Form FCC 601 is a multipurpose form used to apply for an authorization to operate radio stations, amend pending applications, modify existing licenses, and perform a variety of other miscellaneous transactions covered in the Wireless Telecommunications Bureau (WTB) radio services. Form FCC 601 is also a multipart form comprising a main form and several optional schedules. Schedule I is a supplementary schedule used to apply for an authorization to operate a radio station in the categories of fixed microwave and microwave broadcast auxiliary services.

7.2.3 Homologation

One of the biggest headaches facing telecommunications equipment suppliers is not technological but rather regulatory. Equipment is very often sold and installed in countries other than country of manufacture. These countries may have different functionality and safety requirements. To sell their products in other countries, suppliers must go through a *certification process (homologation)*. Quick and efficient approval for the hardware and software in each of the markets usually requires a detailed knowledge of both telecommunications equipment and the local regulatory environment. Such knowledge is typically available from a local company that is intimately familiar with the requirements of the agencies in the countries in which they operate.

Quite often, frequency allocations and channeling plans for wireless services (cellular, PCS, microwave, satellite, and so on) are different in different countries. To answer questions like, "What's the dialing/numbering plan in Nigeria or Croatia?" "What are the specific requirements for a new interface in China?" and "What is the frequency band and channeling plans for the microwave systems in a specific country of interest?" a local presence across the region and broad technical and regulatory expertise are needed.

In some countries, the certification process is simple. If equipment is approved by the U.S. FCC and meets certain minimum requirements, it is considered to be certified. In other countries, government regulations are used to exclude foreign manufacturers from their markets. This is done by requiring suppliers to meet very stringent requirements and standards and quite often to customize their hardware designs. While this is not necessarily technically demanding, it is very time consuming process.

The *CE Mark* is a requirement for all *telecommunications terminal equipment (TTE)* products sold in the European Union, effective January 1, 1996. The CE marking confirms that a product has been tested and meets the essential requirements of the European Telecom Directive to market it throughout the EU. Obtaining the CE Mark allows a product to be sold in EU countries without any further in-country testing.

7.2.4 Other Regulatory Issues

The National Environmental Protection Act (NEPA) requires all federal agencies to implement procedures to make environmental consideration a necessary part of an agency's decision-making process. As a licensing agency, the FCC complies with NEPA by requiring commission licensees and applicants to review their proposed actions for *environmental consequences*. FCC rules implementing the NEPA require the licensee to consider the potential environmental effects from its construction of antenna facilities or structures and to disclose those effects in an environmental assessment (EA), which is filed with the Commission for review. The Commission solicits public comment on the EAs and assists its licensees in working with the appropriate local, state, and federal agencies to reach agreement on the mitigation of potential adverse effects. The filing of an EA is required when a proposed facility may have a significant effect on historic properties.

The National Historic Preservation Act (NHPA), enacted in 1966, is one of the federal environmental statutes implemented in the FCC's NEPA rules. Under the NHPA, federal agencies are required to consider the effects of federal undertakings on historic sites. Commission licensees and applicants must comply with NHPA procedures for proposed facilities that may affect sites that are listed or eligible for listing in the National Register of Historic Places. This process includes consultation with the relevant State Historic Preservation Officer (SHPO) or Tribal Historic Preservation Officer (THPO) to whether the proposed facility may create an adverse effect on an eligible or listed historic property.

The commission requires applicants, licensees, and tower owners *(applicants)* to consider the impact of proposed facilities under the Endangered Species Act (ESA). Applicants must determine whether any proposed facilities may affect listed, threatened, or endangered species or designated critical habitats, or are likely to jeopardize the continued existence of any proposed threatened or endangered species or designated critical habitats. Applicants are also required to notify the FCC and file an environmental assessment if any of these conditions exist. The U.S. Fish and Wildlife Service (FWS) provides information that applicants may find useful regarding compliance with the ESA. In addition, FWS has formulated and published voluntary guidelines for the siting of towers, and these are designed to address potential effects on migratory birds. These guidelines and an accompanying tower site evaluation form are posted at U.S. Fish and Wildlife Service, Bird Issues. According to FWS, the guidelines reflect FWS' judgment of "the most prudent and effective measures for avoiding bird strikes at towers."

7.3 Logistical and Organizational Challenges

Among the key people during microwave network deployment is an *on-site project manager* who is responsible for the day-to-day management of all project deployment efforts. The on-site project manager has responsibility for overseeing and ensuring the success of all on-site activities and fulfilling customer requirements within the assigned field location. All on-site personnel (employees and/or subcontractors) report to the on-site project manager during the microwave-network deployment process. The on-site project manager has the responsibility to ensure that all on-site personnel understand and perform their assignments, to establish on-site working hours and the chain of command, to act as the liaison between the customer and the deployment team, and to serve as the main contact point for the on-site customer representative.

From a logistical perspective, travel and accommodation, project office space, equipment warehousing, communications infrastructure, and so on are also very important parts of the project manager's responsibilities.

7.3.1 Project Controls and Reporting

The project control process will be established and maintained by the project manager through a dedicated *project organization* and the *project plan*. The project plan is a comprehensive, detailed plan that describes the process that will be implemented so as to meet project objectives. It includes schedules, procedures, guidelines, and process flow illustrations that form the basic foundation of the project.

The *project schedule* contains a milestone timetable for project implementation and illustrates the sequence of events, the responsibilities, and the duration necessary for the microwave network to become ready for commercial launch.

The project manager will conduct design review meetings and submit monthly reports with regard to project status, problem areas, the following month's plan, and any unresolved issues.

A project manager usually submits monthly project reports to the customer on technical progress, conformance to the project schedule and/or project plan, equipment delivery status, installation, and test results. This report will be used to state any special requests for schedule changes or changes related to reprioritizing site installations. The monthly report shall include, but is not limited to, project status, problem areas, the following month's plan, and any unresolved issues.

7.3.1.1 Documentation control. Document control will provide the administrative support and control of all project documentation. This will include, but is not limited to, contract change proposals, manual updates, drawing control, specification changes, and so forth. At the end of the project, the following documents should be supplied:

- Technical manuals for the equipment supplied
- One set of interim as-built drawings for each site
- Final as-built drawings following final acceptance
- Factory test results
- Relevant acceptance test plans (ATPs)
- Training manuals (if training is specified)

7.3.1.2 Record keeping. The document used for tracking day-to-day activities on the microwave project deployment is called the *master MW site database*. It contains all the relevant information about the site and the equipment to be installed. In a wireless network, each cell site may initially have more candidates that, over time, will come down to the one final site. It would be waste of time and money to perform a microwave path survey on all candidates, but they all have to be included in the list while the path survey is performed on the final two or three candidates. The final site location will be chosen based on RF, transmission/microwave, and site acquisition feasibility.

Good record keeping is important because, initially, even a small wireless network of 50 cell sites may have 200 to 300 site candidates and the same number of potential microwave hops.

7.3.1.3 Changes and modifications. During the fast-paced build-out phase, things are sometimes done without a written order or without any written trace of who did the work, who ordered it, and why. On a big project, this could amount to a significant amount of money for which, after the fact, no one wants to take responsibility. To avoid the trap of performing the work without a written authorization, a *change notification form,* also called a *change order,* must be issued. It must contain at least the following information:

- Project name
- Change order number
- Description of change
- Reason for change
- Change initiator
- Signed approval

Contractors and consultants must receive written approval for the deliverable prior to starting the new task or making any changes to the previously defined task.

7.3.2 Outsourcing Services

7.3.2.1 About outsourcing. So far, we have discussed the bidding process and hardware procurement. A very similar process is used for choosing the right supplier of services—engineering, installation, testing, project management, and others. It is not unusual to have different providers for microwave equipment, engineering services, and installation services. Due diligence requires that as much attention be paid to the acquisition of services as to the selection of equipment and suppliers.

To ensure that the work is completed according to the highest standards, utilize a company that has established processes and procedures for health and safety, quality, customer service, performance measurement, environmental issues, and training and development. Selecting appropriately qualified contractors, consultants, and engineers usually results in good engineering designs and can significantly reduce a project's life-cycle costs. Rather than merely meeting minimum standards, the services of appropriately qualified engineers and/or engineering consultants can enhance a project's value to clients through rigorous consideration of alternatives, analyses of long-term operating and maintenance costs, and innovative design. It is therefore in the client's best interest to use a qualification-based selection method, which demonstrates the competence of the engineering consultant in the performance of the required engineering services (design, installation, testing, project management, and so forth).

The decision about outsourcing has to be very carefully considered.[5] In the case of network design and planning, the company is granting strategic responsibility to an external company, so contract management and continuous control over the outsourcer are absolutely necessary. In highly innovative corporate settings, the risk in outsourcing is very high, due to the potential for revealing proprietary and sensitive information to outsiders.

7.3.2.2 Turnkey projects. *Turnkey solutions* to large-scale projects can save money. Using a single point of contact to design, build, and commission a telecommunications network involves many obvious benefits, including lower overall costs, reduced cycle time, regulated quality and safety, and simplified administration and project management. By using a single supplier with project-management abilities, it has been proven that overall costs for a project can be reduced by 15 to 20 percent. The theory is that one can reduce cycle time and duplication by using one company to coordinate all aspects of the job, including resource planning (people, equipment, material management), work schedule, and cost management. This will maintain continuity for the project and therefore reduce costs for materials, equipment, and human resources, as it will not involve any downtime or overlap resulting from bad planning.

Clients may either retain comprehensive engineering services from a project's start to finish or develop a work plan for contracting out specific phases of the wireless network project to various parties. They should determine which alternative is appropriate in a particular situation. A preliminary cost estimate will be only as accurate as the defined scope of a project or problem. In cases where very limited preliminary engineering has been undertaken, this estimate will likely reflect the cost of engineering services contemplated in the scope of the project. In cases where more extensive preliminary design has been completed, and the scope of the project has been well defined, this estimate will likely reflect the total project cost.

It is difficult to provide an accurate estimate of the labor cost (services) for the projects in other parts of the world. The engineering team, installation crew, methods, tools, time frame, and test equipment will depend on system configuration, the scope of provision requested of the vendor, the skill level of the labor, and the nature of the technicians and subcontractors available in that country. The availability of construction materials and machines, the weather, site access, transportation, size of system, and the customer's work schedule will affect planning and costing of the services. Even political situation in some countries can affect network build-out. Sometimes, it is difficult to provide even typical information, since the conditions differ greatly on a country-by-country and a project-by-project basis.

There is also a growing popularity among operators for turnkey contracts in which big wireless equipment suppliers take full responsibility for entire projects, including the construction (and sometimes even the initial operations) of new telecom networks.

7.3.2.3 Consultants and contractors. Companies quite often face the challenge of implementing large and/or complex communications system for which, more often than not, they do not have the proper technical and managerial capacity in terms of experience, education, and available work force. Many of these tasks can be given to contracting and/or consulting companies, and there is a significant trend toward outsourcing in the telecommunications industry today. More and more, operators are realizing the benefits of outsourcing services such as network planning, customer care and billing systems, and construction and operations support systems to third parties.

It is important to understand difference between a "contractor" and "consultant" when we talk about microwave engineering process. These two terms are quite often misunderstood and used interchangeably. If we already have a team of engineers and just need an additional engineer to execute the well known and repetitive processes (e.g., cellular RF of the microwave link design), we hire a contractor. If we need someone to help us with a complex problem, find a new and unique solution, or establish a team of people to undertake a type of project for which we have no previous experience, we hire a consultant. A contractor can be a junior-level engineer with two to three years of engineering experience but, from the consultant, we expect many years of relevant engineering experience. Obviously, expectations as well as the pay rates will be significantly different in these two cases.

7.3.3 Time and Resource Management

It is important to realize that, for the microwave links, both sites must be ready for installation and that, for the ring configuration, all the sites comprising the ring have to be ready and available to accept equipment installation before the system can be installed, tested, and commissioned. Large microwave networks usually have ring configuration sites and/or hub sites that have to be completed before the rest of the network can be built. In other words, in most cases, the more important sites (backbone, ring, and hub sites) and the high-capacity microwave links have to be built first so that spur sites and low-capacity systems can be installed afterward.

Some typical times per hop/link required for certain activities during the microwave project are shown in Table 7.1. These times can vary widely from project to project.

TABLE 7.1   Time and Resource Management

Activity	Time required for high-frequency link	People required	Time required for backbone link	People required
MW link planning	2 hr	1	2 hr	1
MW link design	2 hr	1	4 hr	1
Site/path survey	4 hr	2	8 hr	2
Manufacturing process	14–60 days	n/a	30–180 days	n/a
Site acquisition	10–200 days	≥2	10–200 days	≥2
MW link installation, including antennas	8 hr	2	40 hr	4
MW link testing and commissioning	4 hr	2	4 hr	2

Site and path surveys usually take few hours on a short hop in an urban area but sometimes can take up to two to three days on long hops (≥30 mi) in rural areas. Note that testing will last at least 24 hr, plus about 2 hr to set up the equipment and another 2 hr to remove the equipment and analyze the test results.

One of the issues many people tend to forget is microwave radio production time. Of course, it is possible to order two or three hops off the shelf within a few days, but large orders are made to order. For a network of 100 or more microwave hops, it can take a supplier three to six months to produce and deliver—assuming he is not already overwhelmed with another big order. The good thing about this is that such a large number of links can be installed at a rate of only perhaps two to three per day (depending on the number of installation crews deployed), so radio delivery can be staggered with units delivered directly to the sites according to the project installation plan.

Because network design, sales, manufacturing, and installation could be happening in different locations (different countries or even continents), other time issues are involved:

- Manufacturing time
- Manufacturer to port of entry time
- Customs clearance time
- Port of entry to warehouse or directly to the site time

Mobilization issues and resource management are very important issues, and they can be very different and specific during the microwave project. For example, let us say that equipment, engineering, installation, project management, and other costs (turnkey project) were estimated at $100k per hop for 20 backbone microwave hops. The total cost would therefore be $2M for the entire project. At the last moment, client decides to build only 10 hops and use leased lines in the rest of the network. The quick assumption is that the project cost would be half of the initial one, i.e., about $1M. Unfortunately, things are a bit more complicated than that, since the initial quote included certain equipment (including NMS) discounts, which will be reduced for a smaller number of hops. On the other hand, engineering and installation crew's mobilization costs, project management costs, and so forth really change very little with the number of hops. It is therefore reasonable to expect actual cost of $1.1 to 1.2M.

7.3.4 Project Management Tools

A number of different project management software tools are on the market (e.g., Microsoft Project), although consulting and project management companies quite often use their own specialized software tools, which often are customized for the unique requirements of a communications network. They facilitate tracking and scheduling of the network build-outs and serve as an information storage tool that is used for the network build-out programs. The three main areas that these tools need to address are as follows:

- Schedule tracking and reporting at the client, management, and control level
- Collect and provide information related to managing the deployment of a large telecommunications network
- Integrate cost with the schedule to provide cost control and cost forecasting

The scheduling tool must indicate the site development status using the following items:

- Milestones—an operating company/client level for reporting progress
- Activities—a project management/project control level for analyzing and forecasting site design, acquisition, and construction activities
- Tasks—a control level checklist used by the disciplines to identify and provide the status of the detailed work tasks required for site completion

Site information is collected and recorded by these functional areas:

- Transmission/microwave engineering
- Real estate site acquisition and zoning
- Architectural
- Building permitting
- Construction
- Power/Telco (utilities) coordination
- Geotechnical
- Land survey
- Procurement

Report generation is one of the most important features of every project management tool. Usually, there are a number of different reports to choose from, and a number of customized reports are generally also available. Some of the most commonly used reports in microwave network build-out are the *milestones list, activity Gantt chart, activity status report, activity duration report,* and *activity completion count by week.*

7.4 Ethical Issues

7.4.1 Code of Ethics

Licensed engineers in North America are required to adhere to certain ethical principles in conducting their day-to-day activities. The author of this book is a firm believer that not only must the ethical aspects of engineering be applied by the licensed engineers, they also should be taught in schools and colleges as a part of any profession and/or curriculum. One might ask, "What is a *code of ethics?*"

A code of ethics[6] is a basic guide to professional conduct, and it states that

It is the duty of a practitioner to the public, to the practitioner's employer, to the practitioner's clients, to other licensed engineers of the practitioner's profession, and to the practitioner to act at all times with

- Fairness and loyalty to the practitioner's associates, employers, clients, subordinates and employees
- Fidelity to public needs
- Devotion to high ideals of personal honor and professional integrity
- Knowledge of developments in the area of professional engineering relevant to any services that are undertaken, and
- Competence in the performance of any professional engineering services that are undertaken

Through the code of ethics, licensed engineers have a clearly defined duty to society, which is to regard their duty to public welfare as paramount, above their duties to clients or employers. Their duty to employers involves acting as faithful agents or trustees, regarding client information as confidential and avoiding or disclosing conflicts of interest. Their duty to clients means that professional engineers have to disclose immediately any direct or indirect interests that might prejudice (or appear to prejudice) their professional judgment.

Licensed engineers are obligated to give proper credit for engineering work, uphold the principle of adequate compensation for engineering work, and extend the effectiveness of the profession through the interchange of engineering information and experience. As co-workers and supervisors, licensed engineers are required to cooperate on project work and must not review the work of other licensed engineers who are employed by the same company without the other's knowledge, and must not maliciously injure the reputation or business of other practitioners.

7.4.2 Practical Examples of Ethical Dilemmas

We mentioned some of the basic definitions of a *code of ethics,* but the question is, "What does that really mean in everyday life and work?" In these electronic times, it is very easy to copy and paste someone else's work and use it without giving credit to the author of the original document. In addition, one person taking credit for the work of a team of people, and never acknowledging others' contribution to the work and common goal, is very common in today's workplaces. There are a number of examples from engineering practice that illustrate that a particular practice, although legal, may not necessarily be ethical. Sometimes it is hard to distinguish what falls in what category.

For example, what if an engineering company plans to hire consultants for a certain job to be done on their behalf, and they decide that the best person to perform work is the wife of one of the executive directors of the engineering company? If we assume that she is perfectly capable and highly qualified to do the work, is there conflict of interest that may prevent the engineering company from hiring her? On the other hand, by not hiring her, they may be losing the best person for the job. This very common dilemma shows up in different shapes and forms in everyday life and work.

Another example is derived from the microwave engineering field. It is a common knowledge that, in North America (U.S.A., Canada), only qualified and certified people are allowed to climb communications towers. They have to be trained, licensed, and bonded riggers, equipped with safety devices, harness, helmet, boots, and other appropriate equipment to be allowed to climb a tower. Working in a third-world country on a microwave deployment project, the lead engineer notices that the riggers who are installing antennas and cables have no training, and that they are climbing towers barefoot and without any protective and safety devices. What should be done under these circumstances? The engineer could

1. Ignore the situation and be happy with the reduced costs for training, safety devices, and proper wear.

2. Stop the work immediately, notify superiors (called *duty to report*), and continue only after the situation has been rectified.

3. Quit the job and leave the country immediately.

Readers will not have a problem determining the correct course of action. The engineer has to respect the value of human life and make every effort to ensure its protection.

In many cases, companies are concerned not only about human lives but also about potential legal issues and liabilities that come with accidents and the loss of human life. Engineers have to make sure that working conditions are safe and thereby protect not only the interests of the employees but also of the company itself.

7.5 References

1. Wysocki, R. K., et al., *Effective Project Management,* New York: John Wiley and Sons, 1995.
2. Kerzner, H., *Project Management—A Systems Approach to Planning, Scheduling, and Controlling,* 7th ed., New York: John Wiley and Sons, 2001.
3. Clark, M. P., *Networks and Telecommunications—Design and Operations,* 2nd ed., New York: John Wiley & Sons, 1997.

4. Situation Management Systems, Inc., *Managing Negotiation—Selected Readings on Negotiation Skills,* 1996.
5. Mann-Robinson, T. C., *Network Design—Management and Technical Perspectives,* Boca Raton, FL: CRC Press, 1999.
6. Andrews, G., Kemper J., *Canadian Professional Engineering Practice and Ethics,* Toronto, Canada: Saunders College, A Division of Holt, Rinehart and Winston of Canada, Ltd., 1992.

American Cable Stranding

When working on international projects, the information shown in the following table will come in handy. Most of the world expresses wire size in terms of its diameter, while in North America, American Wire Gauge (AWG) is most common unit. AWG is shown below with its exact equivalent value in mm^2 and diameter (mm).

Table A.1 AWG Equivalent in Square Millimeters

AWG number	Cross section (mm^2)	Diameter (mm)	Conductor resistance (Ω/km)
1000 MCM*	507	29.3	0.036
900	456	27.8	0.04
750	380	25.4	0.048
600	304	22.7	0.061
550	279	21.7	0.066
500	253	20.7	0.07
450	228	19.6	0.08
400	203	18.5	0.09
350	177	17.3	0.10
300	152	16.0	0.12
250	127	14.6	0.14
4/0	107.2	11.68	0.18
3/0	85.0	10.40	0.23
2/0	67.4	9.27	0.29
0	53.4	8.25	0.37
1	42.4	7.35	0.47
2	33.6	6.54	0.57
3	26.7	5.83	0.71
4	21.2	5.19	0.91
5	16.8	4.62	1.12

Table A.1 AWG Equivalent in Square Millimeters (Continued)

AWG number	Cross section (mm^2)	Diameter (mm)	Conductor resistance (Ω/km)
6	13.3	4.11	1.44
7	10.6	3.67	1.78
8	8.34	3.26	2.36
9	6.62	2.91	2.77
10	5.26	2.59	3.64
11	4.15	2.30	4.44
12	3.31	2.05	5.41
13	2.63	1.83	7.02
14	2.08	1.63	8.79
15	1.65	1.45	11.2
16	1.31	1.29	14.7
17	1.04	1.15	17.8
18	0.8230	1.0240	23.0
19	0.6530	0.9120	28.3
20	0.5190	0.8120	34.5
21	0.4120	0.7230	44.0
22	0.3240	0.6440	54.8
23	0.2590	0.5730	70.1
24	0.2050	0.5110	89.2
25	0.1630	0.4550	111.0
26	0.1280	0.4050	146.0
27	0.1020	0.3610	176.0
28	0.0804	0.3210	232.0
29	0.0646	0.2860	282.0
30	0.0503	0.2550	350.0
31	0.0400	0.2270	446.0
32	0.0320	0.2020	578.0
33	0.0252	0.1800	710.0
34	0.0200	0.1600	899.0
35	0.0161	0.1430	1125.0
36	0.0123	0.1270	1426.0
37	0.0100	0.1130	1800.0
38	0.00795	0.1010	2255.0
39	0.00632	0.0897	2860.0

*Shown in MCM (circular mills) for larger cross sections; 1 CM = 1 circular mil = 0.0005067 mm^2; 1 MCM = 1000 circular mils = 0.5067 mm^2.

4/0 is also known as 0000; 1 mil = 1/1000 in = 0.0254 mm.

B

Quick RF Reference Sheet

$$\text{Gain or loss (dB)} = 10\log_{10}\frac{P_2}{P_1}$$

where P_1 = input power
P_2 = output power

$$\text{Power (dBm)} = 10\log_{10}\frac{\text{power (mW)}}{1 \text{ mW}}$$

or

$$\text{Power (dBW)} = 10\log_{10}\frac{\text{power (W)}}{1 \text{ W}}$$

$$0 \text{ dBm} = 1 \text{ mW}$$

$$30 \text{ dBm} = 1 \text{ W}$$

$$+30 \text{ dBm} = 0 \text{ dBW}$$

$$-30 \text{ dBW} = 0 \text{ dBm}$$

The signal-to-noise ratio (S/N) is amount by which a signal level exceeds the noise level.

$$\text{SNR (dBm)} = \text{signal level (dBm)} - \text{noise level (dBm)}$$

Effective isotropically radiated power (EIRP) describes the performance of a transmitting system.

$$\text{EIRP (dBW/dBm)} = \text{Tx output power (dBW/dBm)}$$
$$+ \text{ antenna gain (dBi)} - \text{line loss (dB)}$$

Fade margin is an "extra" signal power added to a link to ensure its continued operation if it suffers from signal propagation effects.

FM (dB) = system gain + ant. gain (Tx + Rx) − propagation losses − cable/connector/branching loss (each end added together)

System gain is total gain of the radio system without considering antennas and cables.

System gain (dB) = Tx power − Rx sensitivity

Free-space path loss is the signal energy lost in traversing a path in free space only, with no other obstructions or propagation issues.

$$\text{FSPL (dB)} = 96.6 + 20\log_{10} \text{ (distance in miles)} + 20\log_{10} \text{ (frequency in GHz)}$$

$$\text{FSPL (dB)} = 92.4 + 20\log_{10} \text{ (distance in kilometers)} + 20\log_{10} \text{ (frequency in GHz)}$$

$$\text{Rx level (dBm)} = \text{Tx power} - \text{cable/connector loss 1} + \text{antenna gain 1} - \text{FSPL} + \text{antenna gain 2} - \text{cable/connector loss 2}$$

C

Units of Conversion

1 mile = 1.609 km = 5,280 ft = 63,360 in

1 m = 1.09361 yd = 3.28084 ft = 0.001 km = 6.21371×10^{-4} mi = 39.3701 in

$1 \text{ m}^2 = 1550 \text{ in}^2 = 10.7639 \text{ ft}^2 = 1.19599 \text{ yd}^2$

$1 \text{ m}^3 = 61023.7 \text{ in}^3 = 1.30795 \text{ yd}^3 = 35.3147 \text{ ft}^3$

1 km/hr = 0.277778 m/s = 0.621371 mi/hr = 3.28084 ft/s

1 lb = 0.453592 kg

1 N = 0.1019716 kp = 0.224809 lbf

$1 \text{ N/m}^2 = 10^{-5}$ bar = 1.45038×10^{-4} = 0.0208854 lbf/ft^2 = 0.101972 kp/m^2 = 9.86923×10^{-6} atm

1 kWh = 3.6×10^6 J = 859.845 kcal = 3412.14 Btu

1 hp= 76.0402 kpm/s = 745.700 W

1 W = 0.238846 cal/s = 3.41214 Btu/hr

1 kcal = 4186.8 J = 0.745700 kWhr

$°C = 5/9 \times (°F - 32)$

$°F = (9/5 \times °C) + 32$

Glossary

3G The next or third generation of wireless technology. These networks are specified to operate at a minimum 2 Mbps when stationary and 384 kbps when used at pedestrian user speeds. 3G standards are coordinated through the ITU's IMT-2000 (International Mobile Telecommunications—2000), the European-based UMTS (Universal Mobile Telecommunications System), and the Third Generation Partnership Project (3GPP), a group formed by GSM-supporting standards bodies.

802.11b IEEE standard ratified in 1999 that defines wireless LANs in the 2.4 GHz band.

AAL2 ATM Adaptation Layer 2. AAL2 has been adopted by ITU-T and ATM forum to reduce packing delay. The idea is to multiplex voice packets from several sources into one ATM cell so that the time to fill a cell can be reduced significantly.

AALn ATM Adaptation Layer type n.

ABAM A designation for 22-ga, 110-Ω, plastic insulated, twisted pair Western Electric cable, normally used in a central office.

ADM Add/drop multiplexer.

AGC Automatic gain control. A system that holds the gain and, accordingly, the output of a receiver substantially constant in spite of input-signal amplitude fluctuations.

AIS Alarm indication signal. A signal transmitted to maintain continuity of transmission. The AIS usually is sent to notify the far-end that a transmission fault exists on the line.

ALBO Automatic line build-out.

AMI Alternate mark inversion. A line code in which the signal carrying the binary value alternates between positive and negative polarities.

Angle diversity A technique using multiple antenna beams to receive multipath signals arriving at different angles.

ANSI American National Standards Institute. ANSI is a nonprofit, privately funded membership organization that coordinates the development of voluntary national standards in the United States. ANSI and IEEE standards are often recognized by many government agencies and organizations in both the United States and abroad.

APS Automatic protection switch. Provides a network element with the ability to detect a failed unit/line and switch to the spare one; 1+1 pairs a protection unit/line with each working unit/line; 1+n pairs a protection unit/line for every n working units/lines.

ARIB   Association of Radio Industries and Business (Japan).

ASAE   Adaptive IF slope amplitude equalization.

ASPR   Automatic span powering repeater.

ASTM   American Society for Testing and Materials.

Asynchronous   A type of transmission in which each character is transmitted independently and without reference to a standard clock.

ATDE   Adaptive time domain equalization.

ATM   Asynchronous transfer mode.

ATM Forum   An organization originally founded by a group of vendors and telecommunication companies; this formal standards body comprises various committees responsible for making ATM-related recommendations and producing implementation specifications.

Attenuation   Reduction in signal magnitude or signal loss, usually expressed in decibels.

AWG   American Wire Gauge. A measurement of wire diameter. The lower the AWG number, the larger the wire diameter. Copper phone wiring usually comes in 24 or 26 AWG.

B8ZS   Bipolar eight zero substitution; a line coding scheme.

Backhaul   Portion of the wireless network that carries the wireless calls from cell site radios back to the *mobile switching center (MSC)* and then on to the appropriate service termination points such as the *public switched telephone network (PSTN)* and the Internet.

Bandwidth   The information capacity of a communications resource, usually measured in bits per second for digital transmission and hertz for analog transmission. Also see *narrowband, wideband,* and *broadband.*

Bellcore   Bell Communications Research (called Telcordia now).

BER   Bit error rate.

BERT   Bit error rate test.

BHCA   Busy hour call attempts.

BLER   Block error rate, applicable to a block of data in which one or more bits are in error. BLER = (errored blocks received)/(total blocks sent).

Bluetooth   A radio technology developed by Ericsson and other companies built around a chip that makes it possible to transmit signal over short distances between phones, computers, and other devices without use of wires. Find more information at http://www.bluetooth.com.

BOC   Bell operating company.

BoM   Bill of materials (same as BoQ).

BoQ   Bill of quantity (same as BoM).

bps   Bits per second.

BPV   Bipolar violation; the detection of any isolated error.

Broadband   A classification of the information capacity (bandwidth) of a communication channel. Broadband is generally taken to mean a bandwidth higher than 2 Mbps.

BSI   British Standards Institute.

Canadian Electrical Code (CEC)   Canadian version of the U.S. National Electrical Code (NEC).

Carrier   A telecommunications provider that owns switch equipment.

CAS   Channel associated signaling

CCC   Clear channel capability; usually requires B8ZS line coding on all elements.

CCIR   International Radio Consultative Committee (now ITU-R).

CCITT   Comité Consultatif International Téléphonique et Télégraphique (obsolete term).

CCS   Common channel signaling.

CDMA   Code division multiple access. This code division technology was originally developed for military use over 30 years ago. CDMA is a multiple access technique that uses code sequences as traffic channels within common radio channels—used for cdmaOne (IS-95) and CDMA2000 air interface.

CDMA2000   A third-generation digital air interface technology.

cdmaOne (IS-95)   A second-generation digital air interface technology pioneered by the U.S. firm Qualcomm and further developed in South Korea.

CEPT   Conférence Européenne des Postes et Télécommunications (European Conference of Postal and Telecommunications Administrations).

CFM   Composite fade margin.

Circuit switching   Basic switching process whereby a circuit between two users is opened on demand and maintained for their exclusive use for the duration of the transmission, as opposed to a dedicated circuit that is held open regardless of whether data is being sent.

CO   Central office. The building that contains the switches for a local telephone company.

CODEC   Coder-decoder; converts analog voice to digital and vice versa.

Concatenation   The linking together of various data structures—for example, two bandwidths joined to form a single bandwidth.

CPE   Customer premises equipment.

CRC-n   Cyclic redundancy check − n bits.

Cross-connect (DSX Panel, MDF)   Distribution system equipment used to terminate and administer communication circuits. In a wire cross-connect, jumper wires or patch cords are used to make circuit connections. In an optical cross-connect, fiber patch cords are used. The cross-connect is located in an equipment room, riser closet, or satellite closet.

CSU   Channel service unit; the interface from CPE to the public T1 line.

D4   Fourth-generation digital channel bank.

DACS   Digital access and cross-connect system.

dB   Decibel.

dBdsx   Decibels with respect to the standard level at the DSX-1 cross-connect.

dBm   Decibels (relative to 1 mW).

DC   Direct current.

Decibel (dB)   A measure for comparing two quantities. In common usage, when two quantities have dimensions of power, then their dB ratio is $10\log(Q1/Q2)$. So if $Q1$ is 10 times greater then $Q2$, it is 10 dB greater. If $Q1$ is equal to $Q2$, then it is 0 dB greater.

DEM   Digital elevation model. A digital model describing the elevation of a map area.

DFM   Dispersive fade margin.

Diffraction   A propagation phenomenon that allows radio waves to propagate beyond obstructions via secondary waves created by the obstruction. Classic types of diffractions are *smooth Earth* and *knife-edge*. There is no line of sight between the transmitter and receiver.

Dish antenna   A dish-like antenna used to link communication sites by wireless transmission of voice or data. Also called *microwave antenna* or *microwave dish antenna*.

Diversity   A technique to reduce the effects of fading by using multiple spatially separated antennas (space diversity) to take independent samples of the same signal at the same time. The theory is that the fading in these signals is uncorrelated and that the probability of all samples being below a threshold at a given instant is low. Diversity can also be frequency, angle, hybrid, etc.

DoD   Department of Defense.

DOT   Department of Telecommunications.

DS/CDMA   Direct sequence CDMA.

DS/SS   Direct sequence spread spectrum.

DS0   Digital signal, level 0; 64 kbps [level zero (0)].

DS1   Digital signal, level 1; 1.544 Mbps, the North American standard.

DS3   Digital signal, level 3; 44.736 Mbps, the North American standard.

DSU   Digital service unit.

DSX-1 Digital service cross-connect, level 1; part of the DS1 specification.

DTE Data terminal equipment.

DTED Digital terrain elevation data.

DTMF Dual-tone multifrequency; "touch-tone" dialing.

DWDM Dense wavelength division multiplexing.

EC European Commission.

EFS Error-free seconds.

EIA Electronic Industries Alliance (formerly Electronic Industries Association). It specifies electrical transmission standards, including those used in networking.

EIRP Effective isotropically radiated power in dBW or dBm. The product of the power supplied to the antenna and the antenna gain in a given direction relative to an isotropic antenna. Equal to the transmitted output power minus cable loss plus the transmitting antenna gain.

$$\text{EIRP} = \text{Tx output power (in dBW or dBm)} + \text{antenna gain (dBi)} - \text{line loss (dB)}$$

EMC Electromagnetic compatibility.

Engset The Engset traffic model is used to explore the relationship between the traffic offered to a trunk group, the blocking that will be experienced by that traffic, and the number of lines provided when there are a finite number of sources from which the traffic is generated. It is used to replace the Erlang B traffic model, which tends to overestimate blocking when the ratio of the number of sources to the number of lines is less than ten.

Antenna beamwidth The directivity of a directional antenna. Defined as the angle between two half-power (–3 dB) points on either side of the main lobe of radiation.

EOW Engineer order wire.

ERC European Radiocommunications Committee.

Erlang In telecommunications, an *erlang* is a nondimensional unit with a value between "0" and "1" that indicates how busy a telephone facility is over a period of time (usually one hour). Agner Krarup Erlang (1878–1929) was a Danish mathematician who invented the formula commonly used to forecast telecommunications traffic. The number "1" applied to a particular telephone circuit would indicate busy 100 percent of the time. Erlang B is a calculation for any one of these three factors if you know or can predict the other two:

- Busy hour traffic (BHT), or the number of hours of call traffic during the busiest hour of operation.

- Blocking, or the percentage of calls that are blocked because not enough lines are available.

- Lines or the number of lines in a trunk group.

An extended version of Erlang B allows you to determine the number of people who, when blocked, retry their calls immediately. Most of the common models of capacity assume that calls are either served immediately or are blocked and overflow. In some applications, blocked calls can be delayed and served later, which leads us to Erlang C distribution, where capacity estimation is a function of delay criteria instead of blocking.

It is common for the wireless systems to require GOS, usually specified as a blocking probability using Erlang B formula, of 2 percent in North America and between 1 and 2 percent in Europe.

ERO European Radiocommunications Office.

ERP Effective radiated power (in a given direction). The product of the power supplied to the antenna and its gain relative to a half-wave dipole in a given direction.

ESF Extended super frame, a DS1 framing format of 24 frames.

ETSI European Telecommunications Standards Institute. A body formed by the European Commission in 1988, which included vendors and operators. ETSI's purpose is to define standards that will enable the European market for telecommunications to function as a single market.

Eye diagram A superposition of segments of a received PAM signal displayed on an oscilloscope or similar instrument. The eye diagram is used to assess impairments in the radio channel.

FAA Federal Aviation Administration.

Fade margin In radio communication systems, the difference in decibels between the power level at the receiver under nonfading conditions and the receiver threshold (the point where the signal-to-noise ratio of the receiver is below an acceptable level).

Fading The variation in signal strength from it normal value. Fading is normally negative and can be either fast or slow. It is normally characterized by the distribution of fades, Gaussian, Rician, or Rayleigh.

FAS Frame alignment signal.

FB Framing bit.

FCC Federal Communications Commission. Regulates interstate communications via licenses, rates, tariffs, standards, limitations, and so forth. Commissioners are appointed by the U.S. president. In Canada, the same function is performed by Industry Canada.

FDD Frequency division duplex.

FDL Facility data link; an embedded overhead channel within the ESF format.

FDM Frequency division multiplexing.

FDMA Frequency division multiple access.

FEBE Far-end block error.

FEC Forward error correction. An encoding technique that allows a limited number of errors in digital stream to be corrected based on knowledge of the encoding scheme used.

FERF Far-end receive failure.

FHSS Frequency-hopping spread spectrum.

Fixed wireless or fixed cellular network Also called *wireless local loop (WLL)*. This apparent contradiction in terms signifies a cellular network that is set up to support fixed rather than mobile subscribers. Increasingly used as a fast and economic way to roll out modern telephone services, since it avoids the need for fixed wires.

Frame relay Protocol for packet-switched data communications.

Free-space loss The amount of attenuation of RF energy on an unobstructed path between isotropic antennas. Loss of energy as the RF propagates away from a source.

Frequency diversity The simultaneous use of multiple frequencies to transmit of information. This is a technique used to overcome the effects of multipath fading, since the wavelength for different frequencies result in different and uncorrelated fading characteristics.

Frequency response Attenuation versus frequency or, in other words, a plot of how a circuit or device responds to different frequencies.

Fresnel zones Areas of constructive and destructive interference created when an electromagnetic wave propagating in free space is reflected (multipath) or diffracted as the wave intersects obstacles. Diffraction is caused by the phenomenon that an obstacle in the path of an electromagnetic wave does not cast a completely sharp shadow. Instead, some of the energy in the wave front is bent away from the line of sight path into the shadowed area behind the obstacle.

FSO Free-space optical.

Gain Antenna gain is a measure of directivity. It is defined as the ratio of the radiation intensity in a given direction to the radiation intensity that would be obtained if the power accepted by the antenna were radiated equally in all directions (isotropically). Antenna gain is expressed in dBi.

Geodetic datum A reference that describes the position, orientation, and scale relationships of a reference ellipsoid to the Earth.

GIS Geographic information system. A common name for software tools to acquire and maintain digital geographic data.

GND Ground (0-V).

GoS Grade of service.

Grooming Consolidating or segregating traffic for efficiency.

GSM Global System for Mobile Communications. Originally defined as a pan-European standard for a digital cellular telephone network, created to support cross-border roaming, GSM is now one of the world's main digital wireless standards. It uses TDMA air interfaces and is implemented in 900-MHz, 1800-MHz, and 1900-MHz frequency bands.

Guard band   A set of frequencies or bandwidth used to prevent adjacent systems from interfering with each other. Guard bands are typically used between different types of systems, at the edges of the frequency allocations.

HDR   High data rate system developed by Qualcomm for CDMA 1.9-GHz carriers.

HF   High frequency.

Hz   Hertz, cycles per second.

IEC   Interexchange carrier.

IEEE   Institute of Electrical and Electronics Engineers. Professional organization that defines networking and other standards.

IFM   Interference fade margin.

IMA   Inverse multiplexing for ATM.

IMT-2000   The term used by the International Telecommunications Union for the specification for the projected third-generation wireless services (3G). Formerly referred to as Future Public Land-Mobile Telephone Systems (FPLMTS).

IN   Intelligent network. A capability in the public telecom network environment that allows new services. Also implies a well developed network infrastructure.

Internet   The name given to the worldwide collection of networks and gateways using the TCP/IP protocol that functions as a single virtual network.

IP   Internet Protocol (see also TCP/IP).

IS-2000   CDMA2000 1X.

IS-54–TDMA   The first North American TDMA cellular system interim standard specification. After a number of revisions, it is known today as IS-136.

IS-95   The specification for the CDMA wireless system.

ISDN   Integrated Services Digital Network. A digital public telecommunications network in which multiple services (voice, data, images, and video) can be provided via standard terminal interfaces. Offers 2×64 kbps over the landline network.

ISM   Instructional, Scientific and Medical. Microwave bands (2.4 GHz and 5.8 GHz) that do not require licensing (at least not in the U.S.).

ISO   International Organization for Standardization. ISO is responsible for a wide range of standards, including many that are relevant to networking. Their best-known contribution is the development of the OSI reference model and the OSI protocol suite.

Isotropic antenna   An antenna that radiates in all directions (about a point) with a gain of unity (not a realizable antenna, but a useful concept in antenna theory). Used as a 0-dB gain reference in directivity calculation (gain).

ISTO   Industry Standards and Technology Organization.

ITU   International Telecommunications Union. Based in Geneva, the ITU is an organization of the UN that oversees telecommunications standards around the world.

ITU-R Recommendations   The international technical standards developed by the Radiocommunication Sector (formerly CCIR) of the ITU. They are the result of studies undertaken by Radiocommunication Study Groups on the use of the radio frequency spectrum in terrestrial and space radiocommunication, including the use of satellite orbits and the characteristics and performance of radio systems. The interconnection of radio systems in public communication networks and the performance required for these interconnections are part of the ITU-T Recommendations.

JISC   Japanese Industrial Standards Committee.

kA   Kiloamperes.

kHz   Kilohertz, 1000 cycles per second.

LAN   Local area network.

Land usage   Numerical classification of different surface types, e.g., water, forest, urban, and so on. This item is also referred to as *clutter*.

Latency   The amount of time it takes a packet to travel from source to destination. Together, latency (delay) and bandwidth define the speed and capacity of a network.

LCT   Local craft terminal, a PC with a web browser used for setup and configuration of microwave (or any other) terminal hardware.

Leased line   A permanent telephone connection between two points set up by a telecommunications common carrier. Typically, leased lines are used by businesses to connect geographically distant offices. Unlike normal dial-up connections, a leased line is always active. The fee for the connection is a fixed monthly rate. The primary factors affecting the monthly fee are distance between end points and the bandwidth of the circuit.

LMDS   Local multipoint distribution services in the 28-GHz band.

Local loop   General term for the line from a telephone customer's premises to the telephone company central office (CO).

LOS   Line of sight. A description of an unobstructed radio path or link between the transmitting and receiving antennas of a communications system.

LTE   Line terminal equipment.

LTS   Laser transmission system.

M24   Multiplexer that converts one DS1 line to 24 voice channels for a CO.

M44   Multiplexer that converts one T1 of 44 ADPCM channels into two PCM T1s.

Minimum bending radius   The amount of bend that a fiber or copper cable can withstand before experiencing performance problems.

MMDS   Multichannel multipoint distribution services in the 2.1 and 2.7 GHz bands.

MPI   Multiple path interference.

MTBF   Mean time between failures.

MTSO   Mobile telephone switching office.

Mux   Multiplexer (multiplexing). Combining two or more signals into a single bit stream that can be individually recovered.

Narrowband   A classification of the information capacity or bandwidth of a communication channel. Narrowband is generally taken to mean a bandwidth of 64 kbps or below.

NATE   National Association of Tower Erectors.

NCRP   National Council for Radiation Protection and Measurement.

NEC   National Electrical Code

NEXT   Near-end crosstalk.

NF   Noise figure.

NIST   National Institute of Standards and Technology.

NIU   Network interface unit; test unit installed at the demarcation point.

NNI   Network-to-network interface.

NTIA   National Telecommunications and Information Administration.

NTS   Naval Telecommunications System.

N-WEST   National Wireless Electronic Systems Testbed.

OA&MP   Operations, administration, and maintenance provisioning.

OC   Optical carrier.

OC-1   Optical carrier level 1 (51.84 Mbps).

OC-12   Optical carrier level 12 (622 Mbps).

OC-3   Optical carrier level 3 (155 Mbps).

OC-48   Optical Carrier level 12 (2.4 Gbps).

OEM   Original equipment manufacturer.

OM   Optical multiplexer.

OOF   Out of frame.

Omnidirectional antenna   Antenna that radiates and receives equally in all directions in azimuth.

OOS   Out of synchronization or out of service.

OS   Operating system.

OSHA   Occupational Safety and Health Administration.

OSI Open System Interconnection. A seven-layer architecture model for communications systems developed by ISO and used as a reference model for most network architectures today.

OTDR Optical time domain reflectometer.

Packet switching Packet switching is a method of handling high-volume traffic that allows for efficient sharing of network resources, as packets from different sources can all be sent over the same channel in the bitstream. A packet-switched network breaks up the information into digital packets that are addressed and individually routed and then reassembled in the correct sequence at the destination. These networks allow the medium to be shared, so they are more efficient than circuit-switched networks.

Packetized voice Refers to any means by which voice traffic is split into packets and then transferred to its end destination. This category would include both IP and Internet telephony. Packets are sent separately over available network resources. The packets have headers denoting their place in the message, and when the destination is reached, the message is reassembled. The process is quick but not yet as quick as traditional telephony, resulting in a delay known as *latency*.

PAM Pulse amplitude modulation.

Path loss The amount of loss introduced by the propagation environment between a transmitter and receiver.

PBX Private branch exchange; a private telephone switching system.

PCM Pulse code modulation.

PCN Personal communication network.

PCS Personal communications service. A generic term for a mass-market mobile personal communications service, independent of the technology used to provide it.

PEP Peak envelope power. The average power supplied to the antenna transmission line by a radio transmitter during one radio frequency cycle at the crest of the modulation envelope taken under normal operating conditions.

PL Private line; a leased line, not switched.

Plenum An air-handling space such as that found above drop-ceiling tiles or in raised floors. Also, a fire-code rating for an indoor cable.

POI Point of interface.

POP Point of presence.

POTS Plain old telephone service.

Propagation The process an electromagnetic wave undergoes as it is radiated from the antenna and spreads out across the physical terrain. The *propagation channel* is the physical medium of electromagnetic wave propagation between the transmit and receive antennas, and it includes everything that influences the propagation between the two antennas.

PRS   Primary reference source. The master clocking source in a network.

PSTN   Public switched telephone network The traditional, wired telephone network.

PTT   Post Telephone and Telegraph company. A governmental agency in many countries.

PUC   Public Utilities Commission. A state regulatory body.

QA   Quality assurance.

QoS   Quality of service. A term used to characterize network availability, quality, and reliability. It is also used to designate different classes of ATM service.

QRSS   Quasirandom signal sequence.

Radome   Radio frequency transparent cover used to protect the antenna from wind load, snow and ice, or dust buildup that would otherwise cause excessive mechanical stress on the tower structure and undesirable radiation pattern distortion. Add-on radomes are most often seen on parabolic antennas.

Radiation pattern   The radiation pattern is a graphical representation in either polar or rectangular coordinates of the spatial energy distribution of an antenna.

RBER   Residual bit-error rate.

RBOC   Regional Bell operating company.

Receiver sensitivity   The minimum RF signal power level required at the input of a receiver for certain performance (e.g., BER).

Refraction   Change in direction of propagating radio energy caused by a change in the refractive index, or density, of a medium.

Repeater   A device used to regenerate an optical or electrical signal (of any frequency) to allow an increase in the system length.

RFEM   Radio frequency electromagnetic field.

RFP   Request for proposal.

RFQ   Request for quotation, sent to equipment suppliers and/or service providers to provide price for the specific item that could be hardware and/or services.

RFT   Request for tender.

RJ-11   Standard four-wire connectors for phone lines.

RJ-45   Standard eight-wire connectors for IEEE 802.3 1Base-T networks.

RMS error   Root-mean-square error. It is also called *standard error* or *standard deviation*. It is an error index that describes an average or mean error for observations made under the same (or similar) measurement conditions.

ROI   Return on investment. The time it takes to pay back (or recoup) the money invested in a technology or strategy.

Rollout   Implementing (deploying) a product or a system.

RZ   Return to zero.

SAR   Specific absorption rate. A measure of the rate of energy absorbed by (dissipated in) an incremental mass contained in a volume element of dielectric materials such as biological tissues. SAR is usually expressed in terms of watts per kilogram (W/kg) or milliwatts per gram (mW/g). Guidelines for human exposure to RF fields are based on SAR thresholds where adverse biological effects may occur. When the human body is exposed to an RF field, the SAR experienced is proportional to the squared value of the electric field strength induced in the body.

SCADA   Supervisory control and data; monitor/control of pipelines, railroads, electrical utility networks, and so on.

SCC   Standards Council of Canada.

SD   Space diversity.

SDH   Synchronous digital hierarchy.

SDRM   Sub-rate digital multiplexer.

SF   Superframe format; a DS1 framing format of 12 frames.

Sidelobes   The radiation lobes in any direction other than that of the main lobe.

SLA   Service level agreement. The contract between a provider and a customer that sets the amount of bandwidth and quality of service, among other things.

SLC   Subscriber loop carrier.

Smart Jack   Network interface unit with cross-connecting and monitoring capabilities.

SMDS   Switched multi-megabit digital service.

SMR   Specialized mobile radio.

SNMP   Simple network management protocol, a set of protocols for managing complex networks. SNMP works by sending messages, called *protocol data units (PDUs)*, to different parts of a network. SNMP-compliant devices, called *agents*, store data about themselves in *management information bases (MIBs)* and return this data to the SNMP requesters.

SNR   Signal-to-noise ratio, defined as

$$\text{SNR (dBm)} = \text{signal level (dBm)} - \text{noise level (dBm)}$$

SONET   Synchronous optical network.

Space diversity   A diversity technique widely used in radiocommunications since the very beginning. It consists of two receive antennas physically (spatially) separated to provide uncorrelated receive signals.

SRDM   Sub-rate data multiplexing.

STM   Synchronous transfer mode. An ITU-defined communications method that transmits a group of time division multiplexed (TDM) streams synchronized to a common reference clock. It reserves bandwidth according to a rigid hierarchy, regardless of actual channel usage.

STM–N   Synchronous transfer module—Level N.

Stratum   Level of clock source used to categorize accuracy.

Synchronous   Type of transmission in which the transmission and reception of all data is synchronized by a common clock, and the data is usually transmitted in blocks rather than individual characters.

T1   A 1.544-Mbps transmission standard.

T-BERD   A trade name for T1 bit-error rate tester made by TTC.

TCP/IP   The data protocols used for the Internet.

TDD   Time division duplexing.

TDM   Time division multiplexing.

TDMA   Time division multiple access. A digital transmission technique used for GSM, D-AMPS (IS-136), and PDC air interfaces. D-AMPS in North America is often just called TDMA.

TE   Terminal equipment.

Telcordia Technologies   Formerly Bellcore, an SAIC company, this is the world's largest provider of operations support systems, network software, and consulting and engineering services to the telecommunications industry.

TFM   Thermal fade margin.

TIA   Telecommunications Industry Association. U.S. telecom industry standards body.

TM   Terminal multiplexer.

TMN   Total/Telecommunications Management Network (ITU-TS M.3010).

TND   Transmission (transport) network design

Topographical map   A map that describes the elevation of the terrain, its nature, and built-up areas.

Tx   Transmission.

UAS   Unavailable seconds.

UHF   Ultra-high frequency.

UL   Underwriters Laboratory. A nonprofit laboratory that examines and tests devices, materials, and systems for safety and has began to establish safety standards.

UMTS   Universal mobile telecommunications system.

UNI   User-to-network interface.

U-NII   Unlicensed National Information Infrastructure; 5-GHz microwave band that does not require licensing (at least in the U.S.A.).

Unstructured 1.544- or 2.048-Mbps link   Link without frame, where all bits are used as one single 1.544-Mbps data channel.

UPS   Uninterruptible power supply.

USGS   United States Geological Survey.

UTC   Universal time coordinated; an international time standard. It is the current term for what used to be known as *Greenwich mean time (GMT)*. In-depth information about time and time-related issues can be found at http://tycho.usno.navy.mil.

UTM   Universal transverse mercator.

UTP   Unshielded twisted pair.

VF   Voice frequency.

VHF   Very high frequency.

VRLA   Valve-regulated lead-acid battery.

WAN   Wide area network.

WCDMA   Wideband CDMA (Ericsson-Nokia version of CDMA). A technology for wideband digital radio communications of Internet, multimedia, video, and other capacity-demanding applications. WCDMA has been selected for the third generation of mobile telephone systems in Europe, Japan, and the United States.

WDM   Wavelength division multiplexing.

Wideband   A classification of the information capacity or bandwidth of a communication channel. Wideband is generally taken to mean a bandwidth between 64 kbps and 2 Mbps.

Wireless communication facilities   A land-use facility supporting antennas and microwave dishes that send and/or receive radio frequency signals that provide commercial mobile services, unlicensed wireless services, and common carrier wireless exchange access services. The facilities include structures, towers, and accessory buildings.

Wireless communications   A system that uses radio transmitters and receivers in place of wire lines; when connected to the evolving public switched network, it provides comprehensive telephone service to customers.

WLAN   Wireless local area network. Wireless LAN access is currently available in three formats, depending on the application needed. The three formats are 900 MHz (best range for in-building LANs with a maximum data rate of 1 Mbps), 2.4 GHz (allows for higher data rates of 11 Mbps), and 5 GHz for highest data rate of up to 54 Mbps and least range.

WLL   Wireless local loop.

XDSL   (Generic) digital subscriber line.

XPIC   Adaptive coupling circuit between two orthogonal co-frequency channels or two alternated adjacent channels, on the same link, used to reduce cross-polar interference during adverse propagation conditions.

Zero code suppression   The insertion of a "one" bit to prevent the transmission of eight or more consecutive "zeros." Used to guarantee minimum pulse density.

Index

About the Author

Harvey Lehpamer owns HL Telecom Consulting in San Diego, California, a firm supplying engineering expertise to different clients. He has over 20 years' experience in the planning, design, and deployment of microwave and other transmission networks around the world. In the past, he worked at both Ericsson Wireless Communications, Inc., and Qualcomm, Inc. Mr. Lehpamer is also the author of *Transmission Systems Design Handbook for Wireless Networks* (Artech House, 2002). He can be reached at HL_2@hotmail.com or through his website www. hltelecomconsulting.com.